Transistor Applications

by Richard F. Shea

Consulting Engineer, Schenectady, New York

John Wiley and Sons, Inc., New York · London · Sydney

Copyright © 1964 by John Wiley & Sons, Inc.

All Rights Reserved
This book or any part thereof
must not be reproduced in any form
without the written permission of the publisher.

Library of Congress Catalog Card Number: 64-13222
Printed in the United States of America

Preface

In the more than ten years since the advent of the junction transistor, many books have been published on practically every aspect of transistor theory and applications. The wisdom of publishing still *another* book on such an already well-covered subject may, therefore, logically be questioned. The answer to this question has to be that another book on the subject, to be a worthwhile contribution to the art, must either present new information, or available information in a new way or for a new purpose. This, I submit, is the object of this book. It is not a textbook—most of the theoretical derivations have been deliberately omitted for the sake of conciseness. It is not a handbook—detailed treatments of all the aspects of the subject could not possibly be included in this amount of space. It is, in essence, a concise treatment of the subject, designed specifically for the industrial engineer who wishes to understand the basic theory of transistor applications, yet has neither the time nor the inclination to become expert in the field. It will not, therefore, generally enable the reader to design immediately transistor circuits, although there is considerable design detail included. It will, I trust, enable him to analyze transistor circuits, and thus learn how they work. With this beginning, and with a reasonable amount of supplementary study, the engineer-reader will then be ready to assume the burden of design.

Although not specifically designed as a textbook, this volume can still prove valuable as a supplemental text. I have had the privilege of teaching an industrial transistor course for many years, using as a text *Transistor Circuit Engineering*. In the course of this teaching I found it valuable to develop a set of supplemental notes to help emphasize certain aspects of the subject that were particularly difficult to understand, or, in some cases, to up-date the material or add new angles to the presentation. These notes formed the basis of the material in this book. Thus, this book can provide considerable practical "meat" to the skeleton formed by the usual theoretical textbook.

Additionally, the book has many qualities which can be of considerable value to the engineer who is already conversant with the basic transistor theory. For example, most textbooks, and many handbooks as well, are deficient from the standpoint of ease of utilization. Transistor specifications today almost universally employ a standard symbology and give their values in specific forms, yet all too often the designer working from available texts must convert these published values to some other form which happened to be more attractive to the author. As one aim, therefore, I determined, as far as practicable, to evolve all equations in such form that the reader could insert available specifications directly, without the necessity of conversion.

There are obvious deficiencies in an approach such as this, as well as advantages. The subject coverage is as extensive as is warranted for the avowed purpose, to give the reader understanding, not expertness. In many portions of the book, this means, however, that only typical designs can be included. As an example, the subject of transistor logic can easily be, and has been, expanded to encompass whole volumes. To compress such a subject within the relatively few pages alloted to it here therefore presents quite a challenge. I sincerely believe that I have managed to supply the necessary tools to understand the elements of this subject, and of the other phases of transistor technology included. Of course, having boiled the "fat" off the subject, as it were, the essential information left behind becomes easier to find! On this hopeful note I will end this preface.

RICHARD F. SHEA

October 1963
Schenectady, N.Y.

Contents

Chapter 1   Two-Port Networks 1
  1.1 Introduction 1
  1.2 Network Conventions 1
  1.3 Matrix Representation of Network Equations 3
  1.4 Matrix Determinants 6

Chapter 2   Matrix Interconnections 9
  2.1 Networks in Series 9
  2.2 Networks in Parallel 9
  2.3 Networks in Series-Parallel 10
  2.4 Networks in Parallel-Series 11
  2.5 Cascaded Networks 11
  2.6 Restrictions on Network Interconnections 13
  2.7 Application of Network Interconnections and Matrix Combinations 13

Chapter 3   The Terminated Network 14
  3.1 Input and Output Terminations 14
  3.2 Power Gain 17
  3.3 Transducer Gain 17

Chapter 4   Equivalent Circuits 19
  4.1 Functions of Equivalent Circuits 19
  4.2 Voltage and Current Generator Concepts 19
  4.3 Millman Theorem 20
  4.4 The Z Equivalent Circuit 22
  4.5 The Y Equivalent Circuit 22
  4.6 The H Equivalent Circuit 23

CONTENTS

- 4.7 Equivalent Circuits of the Other Matrix Representations 24
- 4.8 Advantages of the Various Representations 24

Chapter 5   Characteristic Curves   25

- 5.1 Types of Characteristic Curves 25
- 5.2 Transistor Voltage and Current Conventions 25
- 5.3 Typical Characteristics for Germanium Transistors 27
- 5.4 Typical Characteristics for Silicon Transistors 29
- 5.5 Transistor Parameter Nomenclature 33
- 5.6 Relationship Between Parameters and Characteristic Curves 33
- 5.7 Determination of Small-Signal Parameters 34
- 5.8 Specification Values of Transistor Parameters 35
- 5.9 Variation of Parameters with Operating Point and Temperature 36
- 5.10 Static Characteristics for the Unipolar (Field-Effect) Transistor 40

Chapter 6   Transistor Equivalent Circuits   42

- 6.1 The Ideal Transistor 42
- 6.2 Relationship Between the Ideal Parameters and the Transistor Construction 43
- 6.3 The Parameters of the Real Transistor 44
- 6.4 Derivation of the Parameters of the Ideal Transistor from Specification Sheet Values 45
- 6.5 Common-Emitter Equivalent Circuit Using Special Parameters 46
- 6.6 Determination of Network h Parameters from Ideal Transistor Parameters and Parasitic Elements 47
- 6.7 Converting Common-Base Parameters to Common-Emitter and Common-Collector Parameters 48

Chapter 7   Bias   52

- 7.1 Collector Cutoff Current 52
- 7.2 Effect of Variation of V_{BE} 53
- 7.3 Methods of Biasing Transistors 54

7.4 Equations for the Two-Battery Arrangement 55
7.5 Equations for the Single-Battery Arrangement 57
7.6 Graphical Determination of Operating Point 58
7.7 Significance of the Stability Factors 62

Chapter 8 Characteristics of the Single-Stage Amplifier 64

8.1 Introduction 64
8.2 The Equivalent Circuits in Terms of Specification Parameters 65
8.3 Limitations on Use of Equivalent Circuits 67
8.4 Properties of Terminated Common-Base Stage 67
8.5 Properties of Terminated Common-Emitter and Common-Collector Stages 69
8.6 The Degenerated Common-Emitter Stage 70
8.7 Variation of Impedances, Amplification, and Gain with Generator and Load Resistances 71

Chapter 9 Cascaded Stages 76

9.1 Introduction 76
9.2 Basic Relationships in Cascaded Networks 77
9.3 Input Impedance of the Cascaded Pair 77
9.4 Current Amplification of the Cascaded Pair 78
9.5 Voltage Amplification of the Cascaded Pair 79
9.6 Power Gain of the Cascaded Pair 79
9.7 Combinations of Transistor Configurations 79
9.8 The Coupling Network 82
9.9 Effects of the Coupling Capacitor and Emitter By-Pass 85

Chapter 10 Feedback Networks—Matrix Transposition 87

10.1 Introduction 87
10.2 Solution of the Degenerated Emitter Circuit by Matrices 87
10.3 Collector-to-Base Feedback 89
10.4 Use of the Transfer and Transposition Matrices 91
10.5 Analysis of Darlington Pair Using Transposition Matrix 93

Chapter 11   D-C Amplifiers   97

- 11.1   Introduction   97
- 11.2   Direct-Coupled Amplifiers   97
- 11.3   Differential Amplifiers   101
- 11.4   Chopper-Type Amplifiers   104
- 11.5   Carrier-Modulation Amplifiers   106
- 11.6   Zero-Stabilized Amplifiers   107
- 11.7   Field-Effect (Unipolar) Transistors as D-C Amplifiers   108

Chapter 12   Class-A Amplifiers   111

- 12.1   Introduction   111
- 12.2   Small-Signal Class-A Operation   112
- 12.3   Linearity Considerations in Class-A Amplifiers   115
- 12.4   High-Power Class-A Amplifiers   117

Chapter 13   Class-B Amplifiers   121

- 13.1   Introduction   121
- 13.2   Basic Considerations of Class-B Operation   121
- 13.3   Linearity Considerations in Class-B Amplifiers   124
- 13.4   Elimination of Crossover Distortion in Class-B Amplifiers   126
- 13.5   Complementary Class-B Amplifiers   127
- 13.6   Low-Output-Impedance Direct-Drive Amplifier   128

Chapter 14   High-Frequency Parameters   131

- 14.1   Introduction   131
- 14.2   The Basic Parameters   131
- 14.3   Parameters of the Actual Transistor at High Frequencies   132
- 14.4   The High-Frequency Equivalent Circuit   134
- 14.5   Typical Parameter Variation with Frequency   135

Chapter 15   High-Frequency Tuned Amplifiers   140

- 15.1   Introduction   140
- 15.2   The Concept of Normalized Bandwidth   140

15.3 Source and Load Coupled by a
 Single-Tuned Circuit 142
15.4 Double-Tuned Coupling 145
15.5 Transitional Coupling 148
15.6 Procedure for Designing Double-Tuned Circuits 148

Chapter 16 Wide-Band Amplifiers 155

16.1 Introduction 155
16.2 High-Frequency Peaking 155
16.3 Low-Frequency Compensation 161
16.4 Use of Feed-Back in Broad-Band Amplifiers 163
16.5 Distributed Amplifiers 166

Chapter 17 Sinusoidal Oscillators 169

17.1 Introduction 169
17.2 The Oscillator as a Closed-Loop Amplifier 169
17.3 The Colpitts Oscillator 171
17.4 The Hartley Oscillator 174
17.5 The R-C Phase-Shift Oscillator 175
17.6 Other Types of Oscillators 177

Chapter 18 Relaxation Oscillators, Multivibrators 178

18.1 Introduction 178
18.2 The Astable Multivibrator 178
18.3 The Monostable Multivibrator 182
18.4 The Bistable Multivibrator 183
18.5 The Schmitt Trigger 185

Chapter 19 Transient Response 187

19.1 Introduction 187
19.2 Small-Signal Transients 187
19.3 Small-Signal Transient Response,
 Common-Base Configuration 190
19.4 Small-Signal Transient Response,
 Common-Emitter Configuration 191
19.5 Small-Signal Transient Response,
 Common-Collector Configuration 191

19.6 Large-Signal Transient Response, Charge Control Considerations   192

Chapter 20 Negative-Resistance and Switching Devices   199

20.1 Introduction   199
20.2 The Unijunction Transistor   199
20.3 The Four-Layer Diode   204
20.4 Tunnel Diodes   207
20.5 Tunnel Diode Applications   210
20.6 Backward (Back) Diodes   218

Chapter 21 Field-Effect (Unipolar) Transistors   220

21.1 Introduction   220
21.2 Theory of Operation   221
21.3 Static Characteristics of the FET   222
21.4 Biasing the FET   224
21.5 Small-Signal A-C Parameters   225
21.6 Noise   226
21.7 Typical Applications   226

Chapter 22 Logic Circuits   237

22.1 Introduction   237
22.2 The Language of Logic   237
22.3 Logic Elements   241
22.4 Types of Logic Circuits   242
22.5 DCTL Logic   242
22.6 TRL Logic   243
22.7 TDL Logic   245
22.8 ECL Logic   245
22.9 Tolerance Considerations in Logic Circuit Design   246

Chapter 23 Integrated Circuits   249

23.1 Introduction   249
23.2 Types of Integrated Circuits   249
23.3 Integrated Circuit Fabrication   250
23.4 Multiple-Transistor Units   252

23.5 Integrated Logic Circuits 253
23.6 Linear Integrated Circuits 255
23.7 Versatile Custom-Built Assemblies 257
23.8 Limitations on Use of Integrated Circuits 259

Appendix A Symbols   261

Index   269

1
Two-Port Networks

1.1 Introduction

A complicated electronic circuit may be broken up into a number of smaller circuits, or networks, connected in series, parallel, combinations of series and parallel, and cascade. These networks may contain active elements, such as transistors, passive elements, such as resistors or capacitors, or combinations of active and passive elements. Networks containing only passive elements are called passive networks, whereas those containing active elements are called active networks, though they may also contain passive elements.

A network has a number of "ports" to which external connections are made. A two-port network has an input port, to which a source is connected, and an output port, to which a load is usually connected, although occasionally the output may connect to a load containing an active element, such as a voltage source. In transistor circuits the input port is usually supplied from the signal source or a preceding stage, and the output port supplies signal to a load or to a succeeding stage.

1.2 Network Conventions

Figure 1.1 illustrates the conventions usually used in designating input and output currents and voltages. Note that the positive direction for current is flowing *into* the network. If the current actually flows *out* of the network, it is given a negative sign. This convention must be borne in mind when dealing with transistor circuits, since the

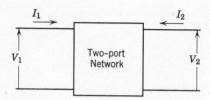

Fig. 1.1 Network voltage and current conventions.

direction of current will depend upon the type of transistor as well as on the circuit details, and we will be frequently encountering negative currents.

It will also be noted that no external connections are shown. This is one of the distinguishing features of network theory, in that it allows us to analyze the network without regard to external connections, then take these into effect subsequently by combining the termination representations with the formulations representing the network.

We can express the relations between the input and output currents and voltages in a number of ways. For example, we can use the impedance parameters and express the network behavior as follows:

$$V_1 = z_{11}I_1 + z_{12}I_2 \qquad (1.1)$$

$$V_2 = z_{21}I_1 + z_{22}I_2 \qquad (1.2)$$

The meaning of the above z parameters may be obtained by assuming that the input or output terminals are respectively open-circuited. Thus, if the output terminal is open I_2 becomes zero, and eqs. 1.1 and 1.2 reduce to $V_1 = z_{11}I_1$ and $V_2 = z_{21}I_1$ respectively. Therefore the parameter z_{11} is evidently the ratio V_1/I_1, or input impedance, with the output open-circuited. Likewise, the parameter z_{21} is V_2/I_1, or the open-circuited output voltage produced by an input current, divided by that input current. This is called the forward transfer impedance. In similar manner, by assuming that the input terminals are open, we can obtain equations defining the other two z parameters:

$$V_1 = z_{12}I_2 \quad \text{and} \quad V_2 = z_{22}I_2$$

Therefore, z_{12} is a backward transfer impedance with the input open-circuited and z_{22} is the output impedance with the input open. Thus we have defined the four z parameters in terms of input and output impedances and transfer impedances with either the input or the output port open.

1.3 Matrix Representation of Network Equations

A matrix equation is, in effect, a shorthand representation of the relationships given in eqs. 1.1 and 1.2. It is written in the following form:

$$\begin{bmatrix} V_1 \\ V_2 \end{bmatrix} = \begin{bmatrix} z_{11} & z_{12} \\ z_{21} & z_{22} \end{bmatrix} \begin{bmatrix} I_1 \\ I_2 \end{bmatrix} \tag{1.3}$$

Each of the three sections of eq. 1.3 contained in brackets is a matrix; thus this equation states that the voltage matrix is the product of an impedance matrix and a current matrix. Equation 1.3 can be further simplified to the form:

$$[V] = [Z][I] \tag{1.4}$$

where each bracketed symbol represents a voltage, impedance, or current matrix as given in eq. 1.3.

Another commonly used set of parameters consists of the admittance parameters. The relationships between inputs and outputs may be shown in the following form:

$$I_1 = y_{11}V_1 + y_{12}V_2 \tag{1.5}$$

$$I_2 = y_{21}V_1 + y_{22}V_2 \tag{1.6}$$

Following the method used for the z parameters, we can obtain meanings for the above y parameters, only now we must short-circuit the input or output terminals to eliminate the V_1 or V_2 terms, as desired. In this manner we find that:

y_{11} = input admittance with the output short-circuited;
y_{12} = backward transfer admittance with input short-circuited;
y_{21} = forward transfer admittance with output short-circuited;
y_{22} = output admittance with input short-circuited.

A third set of parameters, used most extensively in transistor circuit analysis, combines impedance, admittance, and ratios. These are the h parameters, and the current-voltage relationships are:

$$V_1 = h_{11}I_1 + h_{12}V_2 \tag{1.7}$$

$$I_2 = h_{21}I_1 + h_{22}V_2 \tag{1.8}$$

As before, we define these parameters, only now some are defined for short-circuited termination and some for open-circuited termination.

4   TRANSISTOR APPLICATIONS

h_{11} = input impedance with output short-circuited;
h_{12} = backward voltage transfer ratio with input open-circuited;
h_{21} = forward current transfer ratio with output short-circuited;
h_{22} = output admittance with input open-circuited.

A fourth set of parameters also occasionally used in transistor circuit analysis consists of the g parameters, which are expressed as follows:

$$I_1 = g_{11}V_1 + g_{12}I_2 \tag{1.9}$$

$$V_2 = g_{21}V_1 + g_{22}I_2 \tag{1.10}$$

These parameters have the following meanings:

g_{11} = input admittance with output open-circuited;
g_{12} = backward current transfer ratio with input short-circuited;
g_{21} = forward voltage transfer ratio with output open-circuited;
g_{22} = output impedance with input short-circuited.

Finally, there are two other arrangements possible and these use the a and b parameters respectively. The matrix expressions are:

$$\begin{bmatrix} V_1 \\ I_1 \end{bmatrix} = \begin{bmatrix} a_{11} & a_{12} \\ a_{21} & a_{22} \end{bmatrix} \begin{bmatrix} V_2 \\ -I_2 \end{bmatrix} \tag{1.11}$$

$$\begin{bmatrix} V_2 \\ I_2 \end{bmatrix} = \begin{bmatrix} b_{11} & b_{12} \\ b_{21} & b_{22} \end{bmatrix} \begin{bmatrix} V_1 \\ -I_1 \end{bmatrix} \tag{1.12}$$

Note the negative signs preceding the current terms in the third matrix of both equations.

Although these parameters can be defined in similar manner to the other parameters, these definitions can be anomalous, and it is more convenient to define them by relationship to the other, more easily defined, parameters. Thus:

$$a_{11} = \frac{1}{g_{21}} \quad a_{12} = -\frac{1}{y_{21}} \quad a_{21} = \frac{1}{z_{21}} \quad a_{22} = -\frac{1}{h_{21}}$$

and $\tag{1.13}$

$$b_{11} = \frac{1}{h_{12}} \quad b_{12} = -\frac{1}{y_{12}} \quad b_{21} = \frac{1}{z_{12}} \quad b_{22} = -\frac{1}{g_{12}}$$

The above definitions frequently permit calculation of these parameters by inspection. Consider, for example, the simple resistive network of Fig. 1.2. We can obtain all the parameters from their definitions, thus z_{11}, the input impedance with the output open, is obviously $R_1 + R_2$; the output parameter z_{22} is similarly given by $R_2 + R_3$. The transfer impedances z_{12} and z_{21} are given simply by the resistance R_2, since the open-circuit voltages will be $R_2 I_1$ or $R_2 I_2$ when the currents I_1 or I_2 are inserted into their corresponding terminals.

In calculating the y parameters, the input and output terminals are short-circuited; therefore either R_1 or R_3 is shunted across R_2, as the case may be. Thus:

$$y_{11} = \frac{1}{R_1 + R_2 R_3/(R_2 + R_3)}$$

$$y_{12} = \frac{-R_2}{R_1 R_2 + R_1 R_3 + R_2 R_3}$$

$$y_{21} = \frac{-R_2}{R_1 R_2 + R_1 R_3 + R_2 R_3}$$

$$y_{22} = \frac{1}{R_3 + R_1 R_2/(R_1 + R_2)}$$

The negative signs come about because of the current convention previously mentioned. If, for example, an input voltage is applied with the upper terminal positive, the output current flowing through the short circuit will be in a direction away from the network, and thus will have a negative sign according to our convention. The reader is urged to check the foregoing results to obtain a better understanding of the meanings of these parameters.

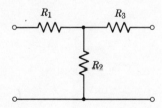

Fig. 1.2 Resistive T network.

Similarly, in our example, the h parameters and g parameters may be found from their definitions:

$$h_{11} = R_1 + \frac{R_2 R_3}{R_2 + R_3} \qquad g_{11} = \frac{1}{R_1 + R_2}$$

$$h_{12} = \frac{R_2}{R_2 + R_3} \qquad g_{12} = -\frac{R_2}{R_1 + R_2}$$

$$h_{21} = -\frac{R_2}{R_2 + R_3} \qquad g_{21} = \frac{R_2}{R_1 + R_2}$$

$$h_{22} = \frac{1}{R_2 + R_3} \qquad g_{22} = R_3 + \frac{R_1 R_2}{R_1 + R_2}$$

The a and b parameters may be found for this circuit from their relationships with the other sets of parameters.

1.4 Matrix Determinants

The determinant of a matrix is designated by means of vertical lines as compared to the brackets used for the matrix representation, thus:

$$\begin{vmatrix} z_{11} & z_{12} \\ z_{21} & z_{22} \end{vmatrix}$$

The symbol for the determinant is the capital delta Δ and a superscript is used to indicate the corresponding parameter set: e.g., Δ^z represents the determinant of the z matrix. The value of the determinant is given by the difference between the products of the diagonal terms, thus:

$$\Delta^z = z_{11}z_{22} - z_{12}z_{21} \tag{1.14}$$

$$\Delta^y = y_{11}y_{22} - y_{12}y_{21} \tag{1.15}$$

$$\Delta^h = h_{11}h_{22} - h_{12}h_{21} \tag{1.16}$$

$$\Delta^g = g_{11}g_{22} - g_{12}g_{21} \tag{1.17}$$

$$\Delta^a = a_{11}a_{22} - a_{12}a_{21} \tag{1.18}$$

$$\Delta^b = b_{11}b_{22} - b_{12}b_{21} \tag{1.19}$$

In similar manner to the parameters themselves, the determinant for one set of parameters may be readily found from those of the other sets. Tables 1.1 and 1.2 present the above relationships between the six different sets of parameters and between their determinants. Thus, referring to Table 1.1 we find that, for example, a_{11} is equal to $-y_{22}/y_{21}$. Thus, in our previous example using the circuit of Fig. 1.2, we could obtain a_{11} as $-(R_1 + R_2)/R_2$. The z determinant Δ^z for this circuit, incidentally, turns out to be the frequently occurring sum $R_1R_2 + R_1R_3 + R_2R_3$.

Table 1.1   Matrix Interrelations

	In Terms of					
	z	y	h	g	a	b
$[z]$	—	$\dfrac{y_{22}}{\Delta^y} \quad \dfrac{-y_{12}}{\Delta^y}$ $\dfrac{-y_{21}}{\Delta^y} \quad \dfrac{y_{11}}{\Delta^y}$	$\dfrac{\Delta^h}{h_{22}} \quad \dfrac{h_{12}}{h_{22}}$ $\dfrac{-h_{21}}{h_{22}} \quad \dfrac{1}{h_{22}}$	$\dfrac{1}{g_{11}} \quad \dfrac{-g_{12}}{g_{11}}$ $\dfrac{g_{21}}{g_{11}} \quad \dfrac{\Delta^g}{g_{11}}$	$\dfrac{a_{11}}{a_{21}} \quad \dfrac{\Delta^a}{a_{21}}$ $\dfrac{1}{a_{21}} \quad \dfrac{a_{22}}{a_{21}}$	$\dfrac{b_{22}}{b_{21}} \quad \dfrac{1}{b_{21}}$ $\dfrac{\Delta^b}{b_{21}} \quad \dfrac{b_{11}}{b_{21}}$
$[y]$	$\dfrac{z_{22}}{\Delta^z} \quad \dfrac{-z_{12}}{\Delta^z}$ $\dfrac{-z_{21}}{\Delta^z} \quad \dfrac{z_{11}}{\Delta^z}$	—	$\dfrac{1}{h_{11}} \quad \dfrac{-h_{12}}{h_{11}}$ $\dfrac{h_{21}}{h_{11}} \quad \dfrac{\Delta^h}{h_{11}}$	$\dfrac{\Delta^g}{g_{22}} \quad \dfrac{g_{12}}{g_{22}}$ $\dfrac{-g_{21}}{g_{22}} \quad \dfrac{1}{g_{22}}$	$\dfrac{a_{22}}{a_{12}} \quad \dfrac{-\Delta^a}{a_{12}}$ $\dfrac{-1}{a_{12}} \quad \dfrac{a_{11}}{a_{12}}$	$\dfrac{b_{11}}{b_{12}} \quad \dfrac{-1}{b_{12}}$ $\dfrac{-\Delta^b}{b_{12}} \quad \dfrac{b_{22}}{b_{12}}$
$[h]$	$\dfrac{\Delta^z}{z_{22}} \quad \dfrac{z_{12}}{z_{22}}$ $\dfrac{-z_{21}}{z_{22}} \quad \dfrac{1}{z_{22}}$	$\dfrac{1}{y_{11}} \quad \dfrac{-y_{12}}{y_{11}}$ $\dfrac{y_{21}}{y_{11}} \quad \dfrac{\Delta^y}{y_{11}}$	—	$\dfrac{g_{22}}{\Delta^g} \quad \dfrac{-g_{12}}{\Delta^g}$ $\dfrac{-g_{21}}{\Delta^g} \quad \dfrac{g_{11}}{\Delta^g}$	$\dfrac{a_{12}}{a_{22}} \quad \dfrac{\Delta^a}{a_{22}}$ $\dfrac{-1}{a_{22}} \quad \dfrac{a_{21}}{a_{22}}$	$\dfrac{b_{12}}{b_{11}} \quad \dfrac{1}{b_{11}}$ $\dfrac{-\Delta^b}{b_{11}} \quad \dfrac{b_{21}}{b_{11}}$
$[g]$	$\dfrac{1}{z_{11}} \quad \dfrac{-z_{12}}{z_{11}}$ $\dfrac{z_{21}}{z_{11}} \quad \dfrac{\Delta^z}{z_{11}}$	$\dfrac{\Delta^y}{y_{22}} \quad \dfrac{y_{12}}{y_{22}}$ $\dfrac{-y_{21}}{y_{22}} \quad \dfrac{1}{y_{22}}$	$\dfrac{h_{22}}{\Delta^h} \quad \dfrac{-h_{12}}{\Delta^h}$ $\dfrac{-h_{21}}{\Delta^h} \quad \dfrac{h_{11}}{\Delta^h}$	—	$\dfrac{a_{21}}{a_{11}} \quad \dfrac{-\Delta^a}{a_{11}}$ $\dfrac{1}{a_{11}} \quad \dfrac{a_{12}}{a_{11}}$	$\dfrac{b_{21}}{b_{22}} \quad \dfrac{-1}{b_{22}}$ $\dfrac{\Delta^b}{b_{22}} \quad \dfrac{b_{12}}{b_{22}}$
$[a]$	$\dfrac{z_{11}}{z_{21}} \quad \dfrac{\Delta^z}{z_{21}}$ $\dfrac{1}{z_{21}} \quad \dfrac{z_{22}}{z_{21}}$	$\dfrac{-y_{22}}{y_{21}} \quad \dfrac{-1}{y_{21}}$ $\dfrac{-\Delta^y}{y_{21}} \quad \dfrac{-y_{11}}{y_{21}}$	$\dfrac{-\Delta^h}{h_{21}} \quad \dfrac{-h_{11}}{h_{21}}$ $\dfrac{-h_{22}}{h_{21}} \quad \dfrac{-1}{h_{21}}$	$\dfrac{1}{g_{21}} \quad \dfrac{g_{22}}{g_{21}}$ $\dfrac{g_{11}}{g_{21}} \quad \dfrac{\Delta^g}{g_{21}}$	—	$\dfrac{b_{22}}{\Delta^b} \quad \dfrac{b_{12}}{\Delta^b}$ $\dfrac{b_{21}}{\Delta^b} \quad \dfrac{b_{11}}{\Delta^b}$
$[b]$	$\dfrac{z_{22}}{z_{12}} \quad \dfrac{\Delta^z}{z_{12}}$ $\dfrac{1}{z_{12}} \quad \dfrac{z_{11}}{z_{12}}$	$\dfrac{-y_{11}}{y_{12}} \quad \dfrac{-1}{y_{12}}$ $\dfrac{-\Delta^y}{y_{12}} \quad \dfrac{-y_{22}}{y_{12}}$	$\dfrac{1}{h_{12}} \quad \dfrac{h_{11}}{h_{12}}$ $\dfrac{h_{22}}{h_{12}} \quad \dfrac{\Delta^h}{h_{12}}$	$\dfrac{-\Delta^g}{g_{12}} \quad \dfrac{-g_{22}}{g_{12}}$ $\dfrac{-g_{11}}{g_{12}} \quad \dfrac{-1}{g_{12}}$	$\dfrac{a_{22}}{\Delta^a} \quad \dfrac{a_{12}}{\Delta^a}$ $\dfrac{a_{21}}{\Delta^a} \quad \dfrac{a_{11}}{\Delta^a}$	—

Table 1.2 Determinant Interrelations

	In Terms of					
	z	y	h	g	a	b
Δ^z	—	$\dfrac{1}{\Delta^y}$	$\dfrac{h_{11}}{h_{22}}$	$\dfrac{g_{22}}{g_{11}}$	$\dfrac{a_{12}}{a_{21}}$	$\dfrac{b_{12}}{b_{21}}$
Δ^y	$\dfrac{1}{\Delta^z}$	—	$\dfrac{h_{22}}{h_{11}}$	$\dfrac{g_{11}}{g_{22}}$	$\dfrac{a_{21}}{a_{12}}$	$\dfrac{b_{21}}{b_{12}}$
Δ^h	$\dfrac{z_{11}}{z_{22}}$	$\dfrac{y_{22}}{y_{11}}$	—	$\dfrac{1}{\Delta^g}$	$\dfrac{a_{11}}{a_{22}}$	$\dfrac{b_{22}}{b_{11}}$
Δ^g	$\dfrac{z_{22}}{z_{11}}$	$\dfrac{y_{11}}{y_{22}}$	$\dfrac{1}{\Delta^h}$	—	$\dfrac{a_{22}}{a_{11}}$	$\dfrac{b_{11}}{b_{22}}$
Δ^a	$\dfrac{z_{12}}{z_{21}}$	$\dfrac{y_{12}}{y_{21}}$	$-\dfrac{h_{12}}{h_{21}}$	$-\dfrac{g_{12}}{g_{21}}$	—	$\dfrac{1}{\Delta^b}$
Δ^b	$\dfrac{z_{21}}{z_{12}}$	$\dfrac{y_{21}}{y_{12}}$	$-\dfrac{h_{21}}{h_{12}}$	$-\dfrac{g_{21}}{g_{12}}$	$\dfrac{1}{\Delta^a}$	—

2

Matrix Interconnections

2.1 Networks in Series

Consider the arrangement of Fig. 2.1, which shows two networks connected in series, i.e., the input current of network A also is the input current of network B, and, similarly, the output currents are equal. These two series-connected networks may be replaced by one equivalent network, corresponding to the dotted lines, which will have z parameters equal to the sums of the z parameter of the two networks, on a term-by-term basis. For the equivalent single network, the z parameters are:

$$z_{11} = z_{11_a} + z_{11_b} \tag{2.1}$$

$$z_{12} = z_{12_a} + z_{12_b} \tag{2.2}$$

$$z_{21} = z_{21_a} + z_{21_b} \tag{2.3}$$

$$z_{22} = z_{22_a} + z_{22_b} \tag{2.4}$$

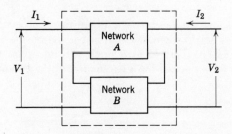

Fig. 2.1 Series connection of networks.

10 TRANSISTOR APPLICATIONS

In the above equations the subscript a denotes the network A, and b the network B. If other than the z parameters are now wanted, they may be obtained by interrelation, as shown in Tables 1.1 and 1.2.

2.2 Networks in Parallel

Figure 2.2 shows two networks connected in parallel, i.e., with the same input and output voltages, but with the input and output currents dividing between the networks. Here the equivalent single network, denoted by the dotted lines, is obtained by adding the y parameters of the two networks on a term-by-term basis:

$$y_{11} = y_{11a} + y_{11b} \tag{2.5}$$

$$y_{12} = y_{12a} + y_{12b} \tag{2.6}$$

$$y_{21} = y_{21a} + y_{21b} \tag{2.7}$$

$$y_{22} = y_{22a} + y_{22b} \tag{2.8}$$

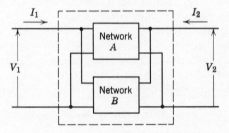

Fig. 2.2 Parallel connection of networks.

2.3 Networks in Series-Parallel

Figure 2.3 shows an arrangement where the two networks are connected in series on the input side and in parallel on the output. This combination may be replaced by a single network whose h parameters are made up of the sums of the h parameters of the two networks:

$$h_{11} = h_{11a} + h_{11b} \tag{2.9}$$

$$h_{12} = h_{12a} + h_{12b} \tag{2.10}$$

$$h_{21} = h_{21a} + h_{21b} \tag{2.11}$$

$$h_{22} = h_{22a} + h_{22b} \tag{2.12}$$

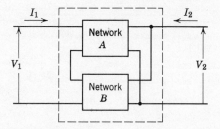

Fig. 2.3   Series-parallel connection of networks.

2.4   Networks in Parallel-Series

In this arrangement the inputs are in parallel and the outputs connected in series, as shown in Fig. 2.4. The equivalent single network is obtained, using the g parameters, by adding the corresponding g parameters of the two networks:

$$g_{11} = g_{11a} + g_{11b} \tag{2.13}$$

$$g_{12} = g_{12a} + g_{12b} \tag{2.14}$$

$$g_{21} = g_{21a} + g_{21b} \tag{2.15}$$

$$g_{22} = g_{22a} + g_{22b} \tag{2.16}$$

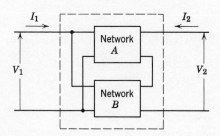

Fig. 2.4   Parallel-series connection of networks.

2.5   Cascaded Networks

In this arrangement, illustrated by Fig. 2.5, the a parameters are used, and the two networks may be replaced by one having a parameters

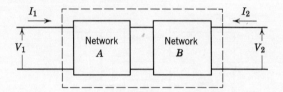

Fig. 2.5 Cascaded networks.

which are obtained by multiplying the a matrices rather than adding them. Now the a matrix terms are:

$$a_{11} = a_{11_a}a_{11_b} + a_{12_a}a_{21_b} \tag{2.17}$$

$$a_{12} = a_{11_a}a_{12_b} + a_{12_a}a_{22_b} \tag{2.18}$$

$$a_{21} = a_{21_a}a_{11_b} + a_{22_a}a_{21_b} \tag{2.19}$$

$$a_{22} = a_{21_a}a_{12_b} + a_{22_a}a_{22_b} \tag{2.20}$$

Frequently, the parameters of the two networks connected in cascade are given in one of the more common forms, e.g., the z parameters. Under these conditions it is tedious to convert from the z form to the corresponding a parameters and then perform the above multiplication and then convert back again to some other form. For the three more common forms of parameters, the following equations may be used to obtain the z, y, or h parameters of the equivalent single network from the z, y, or h parameters of the two component networks.

$$z_{11} = \frac{z_{11_a}z_{11_b} + \Delta^z_a}{z_{22_a} + z_{11_b}} \tag{2.21}$$

$$z_{12} = \frac{(z_{11_a}z_{11_b} + \Delta^z_a)(z_{22_a}z_{22_b} + \Delta^z_b)}{z_{21_a}z_{21_b}(z_{22_a} + z_{11_b})} - \frac{z_{11_a}\Delta^z_b + \Delta^z_a z_{22_b}}{z_{21_a}z_{21_b}} \tag{2.22}$$

$$z_{21} = \frac{z_{21_a}z_{21_b}}{z_{22_a} + z_{11_b}} \tag{2.23}$$

$$z_{22} = \frac{z_{22_a}z_{22_b} + \Delta^z_b}{z_{22_a} + z_{11_b}} \tag{2.24}$$

$$y_{11} = \frac{y_{11_a}y_{11_b} + \Delta^y_a}{y_{22_a} + y_{11_b}} \tag{2.25}$$

$$y_{12} = \frac{y_{11_a}\Delta^y_b + \Delta^y_a y_{22_b}}{y_{21_a}y_{21_b}} - \frac{(y_{11_a}y_{11_b} + \Delta^y_a)(y_{22_a}y_{22_b} + \Delta^y_b)}{y_{21_a}y_{21_b}(y_{22_a} + y_{11_b})} \tag{2.26}$$

$$y_{21} = -\frac{y_{21_a}y_{21_b}}{y_{22_a} + y_{11_b}} \qquad (2.27)$$

$$y_{22} = \frac{y_{22_a}y_{22_b} + \Delta^y_b}{y_{22_a} + y_{11_b}} \qquad (2.28)$$

$$h_{11} = \frac{h_{11_a} + \Delta^h_a h_{11_b}}{1 + h_{22_a}h_{11_b}} \qquad (2.29)$$

$$h_{12} = \frac{h_{11_a}h_{22_b} + \Delta^h_a \Delta^h_b}{h_{21_a}h_{21_b}} - \frac{(h_{11_a} + \Delta^h_a h_{11_b})(h_{22_a}\Delta^h_b + h_{22_b})}{h_{21_a}h_{21_b}(1 + h_{22_a}h_{11_b})} \qquad (2.30)$$

$$h_{21} = -\frac{h_{21_a}h_{21_b}}{1 + h_{22_a}h_{11_b}} \qquad (2.31)$$

$$h_{22} = \frac{h_{22_a}\Delta^h_b + h_{22_b}}{1 + h_{22_a}h_{11_b}} \qquad (2.32)$$

2.6 Restrictions on Network Interconnections

The above matrix manipulations are permissible provided that the interconnection does not change the operation of either individual network. The input and output currents of the two networks must be unchanged and, in particular, care must be exercised to avoid short-circuiting any internal elements of one network by the interconnection with the other.

2.7 Application of Network Interconnections and Matrix Combinations

The preceding combinations can prove extremely useful in analyzing feedback arrangements, for example, the single-stage with either series feedback by virtue of an unbypassed emitter resistor or an output-to-input feedback path. It will be shown in a subsequent chapter that these arrangements can be easily analyzed by use of the z and y matrix representations of the transistor and feedback elements respectively.

The cascaded network equations are useful in analyzing certain complex feedback circuits where it becomes necessary to transpose the circuit to obtain a workable arrangement from the analysis standpoint. Another example of the utilization of the cascaded-network equations is in the combination of our active network and a passive coupling network.

3

The Terminated Network

3.1 Input and Output Terminations

We have shown how any network may be analyzed to derive descriptive parameters by using any of the six sets described in Chapter 1. In this chapter the modifications due to the use of other than open- or short-circuited terminations will be developed, and equations will be presented for obtaining the effects of practical terminations.

Figure 3.1 shows a network being supplied from a generator which has an internal voltage V_g and source impedance Z_g. A load impedance Z_l is connected to the output terminals. The input and output currents and voltages correspond to the previously shown network conventions.

As shown in Fig. 3.2, the network can be considered the equivalent of an input impedance Z_i, to which the source is connected, and an internal generator of voltage V_o, supplying the output voltage V_2 through the internal output impedance Z_o. Alternatively, as shown in

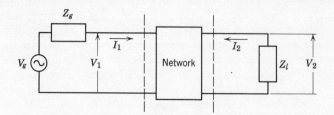

Fig. 3.1 The terminated network.

THE TERMINATED NETWORK 15

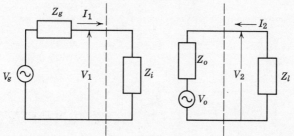

Fig. 3.2 Equivalent circuit representation of terminated network.

Fig. 3.3, the output can be considered to be a current generator I_o, supplying current to an internal admittance Y_o and the load admittance Y_l.

In addition to the input and output impedances, the current and voltage amplifications and power gains are of interest. The current amplification is the ratio I_2/I_1, and the voltage amplification is the ratio V_2/V_1. These ratios may be pure numerics or, at high frequencies, become complex numbers, indicating phase shift between input and output currents or voltages. Power gain may be indicated in a number of ways and will be treated in a later section of this chapter.

Table 3.1 presents the equations for input and output impedances and current and voltage amplification, using any of the six sets of network parameters. For example, the input impedance is given as

$$Z_i = \frac{\Delta^z + z_{11}Z_l}{z_{22} + Z_l}$$

This equation can be used to indicate the effect of output load on input impedance. As the load approaches zero, its effect vanishes and the input impedance approaches Δ^z/z_{22}. Reference to Table 1.1 shows that

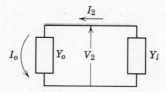

Fig. 3.3 Alternative output configuration.

Table 3.1 Properties of the Terminated Network

Z_i	$\dfrac{\Delta^z + z_{11}Z_l}{z_{22} + Z_l}$	$\dfrac{1 + y_{22}Z_l}{y_{11} + \Delta^y Z_l}$	$\dfrac{h_{11} + \Delta^h Z_l}{1 + h_{22}Z_l}$	$\dfrac{g_{22} + Z_l}{\Delta^g + g_{11}Z_l}$	$\dfrac{a_{12} + a_{11}Z_l}{a_{22} + a_{21}Z_l}$	$\dfrac{b_{12} + b_{22}Z_l}{b_{11} + b_{21}Z_l}$
Z_o	$\dfrac{\Delta^z + z_{22}Z_g}{z_{11} + Z_g}$	$\dfrac{1 + y_{11}Z_g}{y_{22} + \Delta^y Z_g}$	$\dfrac{h_{11} + Z_g}{\Delta^h + h_{22}Z_g}$	$\dfrac{g_{22} + \Delta^g Z_g}{1 + g_{11}Z_g}$	$\dfrac{a_{12} + a_{22}Z_g}{a_{11} + a_{21}Z_g}$	$\dfrac{b_{12} + b_{11}Z_g}{b_{22} + b_{21}Z_g}$
$\dfrac{V_2}{V_1}$	$\dfrac{z_{21}Z_l}{\Delta^z + z_{11}Z_l}$	$\dfrac{-y_{21}Z_l}{1 + y_{22}Z_l}$	$\dfrac{-h_{21}Z_l}{h_{11} + \Delta^h Z_l}$	$\dfrac{g_{21}Z_l}{g_{22} + Z_l}$	$\dfrac{Z_l}{a_{12} + a_{11}Z_l}$	$\dfrac{\Delta^b Z_l}{b_{12} + b_{22}Z_l}$
$\dfrac{I_2}{I_1}$	$\dfrac{-z_{21}}{z_{22} + Z_l}$	$\dfrac{y_{21}}{y_{11} + \Delta^y Z_l}$	$\dfrac{h_{21}}{1 + h_{22}Z_l}$	$\dfrac{-g_{21}}{\Delta^g + g_{11}Z_l}$	$\dfrac{-1}{a_{22} + a_{21}Z_l}$	$\dfrac{-\Delta^b}{b_{11} + b_{12}Z_l}$

this is the same as h_{11}, which has been defined as the short-circuited input impedance; thus we check our definition and the equations. As Z_l approaches infinity, Z_i approaches z_{11}, which was defined as the open-circuited input impedance, and and again this agrees with the equations.

3.2 Power Gain

The power gain of a network is the ratio of the power delivered to the load to that delivered to the input of the network. Referring to Fig. 3.1, the output power is $R_l(I_2)^2$, where R_l is the real part of the load impedance. The input power is $R_i(I_1)^2$, where R_i is the real part of the input impedance. The power gain is, therefore, $(A_i)^2 R_l/R_i$, where A_i is the current amplification I_2/I_1. If Z_l is complex, it must be inserted in the equation for input impedance and the real part of that impedance must be determined and used in the above equation. If the load is nonreactive, the equation for power gain becomes much simpler, reducing, for example, to the following by using the h parameters:

$$G = \frac{(h_{21})^2 R_l}{(1 + h_{22}R_l)(h_{11} + \Delta^h R_l)} \qquad (3.1)$$

3.3 Transducer Gain

Transducer gain is the ratio of the power delivered to the load to that available from the generator, and is usually of more interest than the above form of power gain, since it takes into consideration the effect of input mismatch. Again, by using the h parameters, which are the most commonly used in transistor circuit design, the transducer gain is given by the following equation, for the case where the load is nonreactive and the transistor does not produce a reactive component of input impedance, i.e., at relatively low frequencies.

$$G_t = \frac{4 R_g R_l h_{21}^2}{[R_g(1 + h_{22}R_l) + h_{11} + \Delta^h R_l]^2} \qquad (3.2)$$

REFERENCES

1. Kraus, A. D., and G. F. Ross, "Determinants and Matrices," *Electrical Manufacturing* (Dec. 1959), pp. 133–148.

2. Weinberg, L., "Network Analysis," *Electrical Manufacturing* (Jan. 1960), pp. 89–116.
3. Weinberg, L., "An Introduction to Network Synthesis," *Electro-Technology* (Jan. 1961), pp. 85–104.
4. LePage, W. R. and S. Seely, *General Network Analysis*, McGraw-Hill, New York, 1952, pp. 139–173.
5. Linvill, J. G. and J. F. Gibbons, *Transistors and Active Circuits*, McGraw-Hill, 1961, pp. 209–260.
6. Shea, R. F. et al., *Principles of Transistor Circuits*, Wiley, New York, 1953, pp. 301–338, 509–513.
7. Getgen, L. E., "Application of Matrix Algebra to Circuit Design," *Electro-Technology* (Feb. 1963), pp. 70–79.

4

Equivalent Circuits

4.1 Functions of Equivalent Circuits

An equivalent circuit may be used in a number of ways. Given a number of linear equations representing the relationship between the various input and output currents and voltages, it is frequently desirable to be able to derive a circuit which will perform in identical manner, even though the components of this circuit may not be the same as those in the actual circuit; in some cases the values of the actual circuit are not conveniently ascertained. An equivalent circuit frequently allows intuitive insight into the behavior of the overall circuit into which the network is inserted. Additionally, if the relationship between various matrix forms and their equivalent circuits can be established, it is possible to determine the matrix parameters of a particular circuit frequently by the process of rearranging the circuit elements in the form of one of these equivalent circuits.

4.2 Voltage and Current Generator Concepts

Voltage and current generators are frequently used in equivalent circuits, in fact, a circuit cannot be active without one or more such generators. Figure 4.1 shows the symbols used for voltage and current generators in this book. In Fig. 4.1a the voltage generator is shown as a circle, with voltage V_g. Z_g is the associated generator impedance. The open circuit voltage will be V_g and the source impedance, looking back into the terminals 1, 2 will be Z_g. In Fig. 4.1b the equivalent current

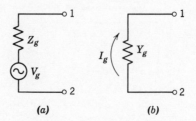

Fig. 4.1 Voltage and current generator conventions.

generator is designated by the curved arrow. This indicates that a current of this value flows from an infinite impedance source into the points indicated by the arrow, and will divide between the source admittance Y_g and the external load in proportion to their admittances. I_g is the current that will flow into an external short circuit and Y_g is the source admittance, looking back into the terminals 1, 2.

A voltage generator concept can be converted to the equivalent current generator by the following relationships:

$$Y_g = \frac{1}{Z_g} \tag{4.1}$$

$$I_g = \frac{V_g}{Z_g} = V_g Y_g \tag{4.2}$$

4.3 Millman Theorem

The use of this theorem allows relatively straightforward determination of branch currents in a somewhat complex network, often merely by inspection. The utilization of this method is illustrated by Fig. 4.2, where point O is any point in the overall circuit of which the three impedances Z_1, Z_2, and Z_3 are parts, and between which point and the points 1, 2, and 3 the voltages V_{O1}, V_{O2}, and V_{O3} are known. The equation relating these voltages and impedances is:

$$V_{OO'} = \frac{V_{O1}Y_1 + V_{O2}Y_2 + V_{O3}Y_3}{Y_1 + Y_2 + Y_3} \tag{4.3}$$

In general terms, the voltage between the two points O and O' is the sum of the products of the branch voltages and admittances, divided by

EQUIVALENT CIRCUITS 21

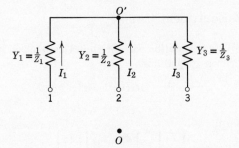

Fig. 4.2 Circuit illustrating Millman theorem.

the sum of all the admittances. This theorem can, naturally, be extended to any number of branches and is not limited to the three shown here.

As an example of the application of this theorem, consider the circuit shown in Fig. 4.3, where voltages exist in two of the branches, but not in the third, and it is desired to determine the current flowing in this latter branch. Applying eq. 4.3,

$$V_{OO'} = \frac{-V_1 Y_1 - V_3 Y_3}{Y_1 + Y_2 + Y_3} \qquad (4.4)$$

whence

$$I_2 = \frac{(V_1 Y_1 + V_3 Y_3) Y_2}{Y_1 + Y_2 + Y_3} \qquad (4.5)$$

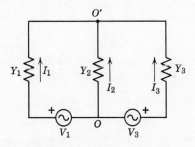

Fig. 4.3 Example of application of Millman theorem.

4.4 The Z Equivalent Circuit

The Z matrix representation was shown in Chapter 1 to be the equivalent of the following equations:

$$V_1 = z_{11}I_1 + z_{12}I_2 \tag{4.6}$$

$$V_2 = z_{21}I_1 + z_{22}I_2 \tag{4.7}$$

or in matrix form:

$$\begin{bmatrix} V_1 \\ V_2 \end{bmatrix} = \begin{bmatrix} z_{11} & z_{12} \\ z_{21} & z_{22} \end{bmatrix} \begin{bmatrix} I_1 \\ I_2 \end{bmatrix} \tag{4.8}$$

These equations can be represented by the equivalent circuit shown in Fig. 4.4, where each mesh consists of one voltage generator in series with an impedance. Comparison of this circuit with the definitions of the z parameters given earlier will prove the equivalence, e.g., z_{11} is obviously the input impedance with the output open-circuited, since now the current I_2 does not exist and the voltage generator therefore does not exist. The polarity signs on the generators imply that the generators will have this polarity if the currents are in the direction shown.

4.5 The Y Equivalent Circuit

The equations for the Y matrix are:

$$I_1 = y_{11}V_1 + y_{12}V_2 \tag{4.9}$$

$$I_2 = y_{21}V_1 + y_{22}V_2 \tag{4.10}$$

or, in matrix form,

$$\begin{bmatrix} I_1 \\ I_2 \end{bmatrix} = \begin{bmatrix} y_{11} & y_{12} \\ y_{21} & y_{22} \end{bmatrix} \begin{bmatrix} V_1 \\ V_2 \end{bmatrix} \tag{4.11}$$

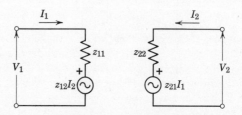

Fig. 4.4 Z equivalent circuit.

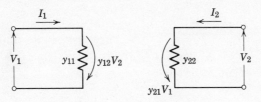

Fig. 4.5 Y equivalent circuit.

These equations are represented by the equivalent circuit shown in Fig. 4.5, where the meshes are now represented by admittances shunted by current generators. The directions of the arrows on the current generators correspond to the upper ends of the voltages on input and output being positive with respect to the lower ends. Again, this circuit may be checked against the definitions of the y parameters; thus, y_{11} is obviously the input admittance with the output short-circuited, since under this condition the output voltage is zero and the current generator does not exist.

4.6 The H Equivalent Circuit

The equations for the h parameters are:

$$V_1 = h_{11}I_1 + h_{12}V_2 \tag{4.12}$$

$$I_2 = h_{21}I_1 + h_{22}V_2 \tag{4.13}$$

or in matrix form,

$$\begin{bmatrix} V_1 \\ I_2 \end{bmatrix} = \begin{bmatrix} h_{11} & h_{12} \\ h_{21} & h_{22} \end{bmatrix} \begin{bmatrix} I_1 \\ V_2 \end{bmatrix} \tag{4.14}$$

Figure 4.6 satisfies these equations, hence, is the equivalent circuit for the H matrix. It will be noted that this circuit now mixes current and voltage generators, impedance, and admittance. There is a voltage generator in the input mesh which has the polarity shown when the output voltage V_2 has its upper end positive with respect to the lower. The current generator in the output mesh will have the direction shown when the input current flows into the circuit as shown.

24 TRANSISTOR APPLICATIONS

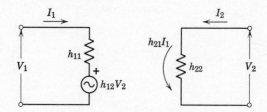

Fig. 4.6   *H* equivalent circuit.

4.7 Equivalent Circuits of the Other Matrix Representations

An approach similar to that used above will serve to derive the equivalent circuits for the other three matrix representations. It will be found that these are composed of similar combinations of voltage and current generators and impedances and admittances, related to those shown in the above mentioned three circuits. By using the table of matrix interrelationships, it is easy to obtain the values of the other parameters for substitution in their equivalent circuits.

4.8 Advantages of the Various Representations

The relative advantages of the various parameter representations lie more in the practical considerations of measurement than in any other factors. For example, the z parameters are open-circuit parameters. In the case of transistors, it becomes extremely difficult to make measurements of these parameters, since it is virtually impossible to obtain terminations which will permit application of the required d-c supply voltages and yet present essentially infinite impedance at the measurement frequency. The y parameters are better, but even here there is some difficulty in obtaining short-circuited input terminations, since the transistor input impedance is so low. For these reasons, the h parameters are most generally used in transistor design, and are commonly given on the specification sheets, as we will see in a subsequent chapter. Under some conditions, however, the other representations offer distinct advantages, and it becomes desirable to convert to them.

5
Characteristic Curves

5.1 Types of Characteristic Curves

In the two-port network of Fig. 1.1, we had two voltages and two currents. A characteristic curve is a plot of one of these as a function of the others. Obviously there are many ways in which these relationships can be shown; for example we can apply a fixed input voltage V_1 to the network and vary the output voltage V_2, noting the variation of the output current I_2. We can then adjust V_1 to a new value and run a new plot of V_2 versus I_2, and by repeating the process, secure a family of curves covering the useful range of these variables. In order to find I_1, we must now plot a new set of curves in which I_1 is found as a function of one of the other variables, say V_1, as another variable, say I_2, is held constant. Thus it becomes evident that the properties of the network of Fig. 1.1 can be determined from any two sets of characteristic plots, and that any of the four variables can be determined from knowledge of the other three.

5.2 Transistor Voltage and Current Conventions

The transistor is a three-element device, like the electron tube, and may be connected with any of the three as the common element. These are designated the common-base configuration, the common-emitter configuration, and the common-collector configuration, respectively. Figure 5.1 illustrates these three configurations for a p-n-p transistor, whereas Fig. 5.2 corresponds to an n-p-n transistor. The directions of

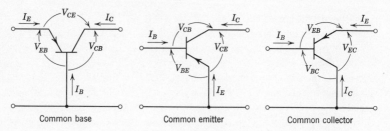

Fig. 5.1 The three configurations for a p-n-p transistor.

currents and voltages are the same for all types under the system of symbols adopted by the industry, despite the fact that the directions are reversed in actual practice. This is accomplished by the convention of using a negative sign where the direction of current or voltage is actually opposite to that shown. Thus, in the case of the p-n-p transistor, the collector current is negative, whereas in an n-p-n transistor it is positive. Conversely, the emitter currents are positive and negative respectively. The emitter-base voltage V_{EB} is positive for a p-n-p, negative for an n-p-n. It should also be noted that the input voltage for the common-emitter configuration, V_{BE}, is the reverse of that for the common-base configuration, hence has the opposite sign. Another difference between configurations is that the collector voltage is specified with respect to the base, i.e., V_{CB} for the common-base configuration, while it is stated with respect to the emitter V_{CE} in the other configuration. In the common-collector configuration, the input voltage is V_{BC} and the output voltage is V_{EC}.

Voltages must always carry two subscripts at least, the first to designate the element from which the voltage is measured, e.g., C for collector,

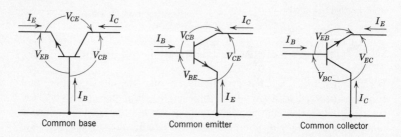

Fig. 5.2 The three configurations for an n-p-n transistor.

the other to designate the reference electrode, e.g., B for base, thus V_{CB} means the voltage from collector to base, and is positive if the collector is positive with respect to the base. Currents will require only one subscript unless a special connotation is implied. Thus I_{CBO} implies the current from the collector to the base, with the emitter open. When such designations are used, the first two letters of the subscript indicate the direction of current, in this case from collector to base, while the third letter indicates the state of the third electrode—O indicating open, S short-circuited, and R loaded with a resistance of value R.

Transistor characteristic curves are commonly plotted in either the common-base or common-emitter configuration, with the latter becoming the more commonly used. The two sets of curves are: (1) the input family, which plots base-to-emitter voltage against emitter, collector, or base current, with the collector-to-base or collector-to-emitter voltage as the fixed control, and (2) the collector family, plotting collector current versus collector-to-base or collector-to-emitter voltage, for fixed values of emitter current or base current respectively.

5.3 Typical Characteristics for Germanium Transistors

Figure 5.3 shows a typical input curve, V_{BE} versus I_C, for a type 2N43 germanium p-n-p transistor. Although this shows only one curve, the effect of the output is not very pronounced. It will be noted that this curve has essentially the typical exponential shape of the

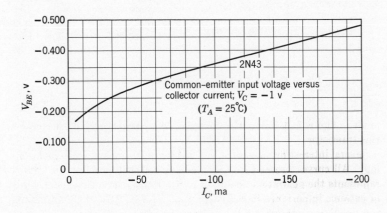

Fig. 5.3 Input characteristic, 2N43 transistor.

28   TRANSISTOR APPLICATIONS

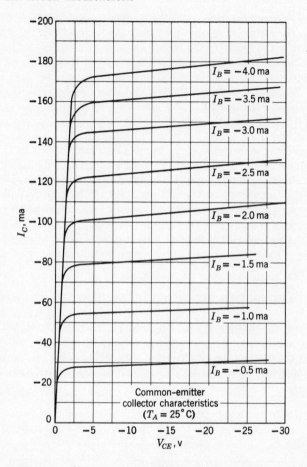

Fig. 5.4 Collector characteristics, 2N43 transistor.

diode. Figure 5.4 shows the collector or output family of curves for this transistor.

Several interesting aspects of this last set of curves are worth mentioning. All curves proceed from one common nearly vertical line, which represents the saturation condition of the transistor. This condition is of extreme importance in switching applications. The slope represents a saturation resistance, and the distance from any point on this line to the vertical axis is the saturation voltage, which is the drop remaining across the transistor when it is driven into saturation. The less this saturation voltage, the better the transistor is as a switch, and the less will be the dissipation in this state. Next, the flatness of the curves above their knees should be noted, as this indicates high output impedance. The spacing between curves is not as uniform as would be the case for the common-base characteristics, indicating some nonlinear relationship between base current and collector current, particularly at high currents.

5.4 Typical Characteristics for Silicon Transistors

Figure 5.5 shows typical input curves for the type 2N1613 silicon transistor and also the variation of this characteristic with temperature. In (a) we see the curve for $-55°C$, (b) shows the curve at room temperature, 25°C, and (c) shows it at 100°C. It will be noted that the shape of the curve is essentially unchanged as temperature varies; however the curves shift toward a lower voltage as temperature rises, for example, V_{BE} drops about 0.2 v as the temperature changes from $-55°$ to $100°$. This shift in base-emitter voltage is of considerable importance in the design of d-c amplifiers, as it is one of the major potential causes of drift with temperature. It will also be noted that these input curves drop to the zero axis at much higher values of voltage than was the case for the germanium transistor; this implies that silicon transistors have a higher cutoff voltage than germanium transistors, and that they can become cut off with quite considerable voltage remaining on the base.

Figures 5.6 and 5.7 show the collector families for the 2N1613 transistor, the former for relatively high voltages and low currents, the latter for high currents and low voltages. As before, the variation with temperature is shown—(a) being for $-55°C$, (b) for room temperature, and (c) for 100°C.

Several details of these curves warrant attention; there is a change in the slope with current, also at high voltages the curves turn upward, and this implies a change of output impedance with current and an overload limit. All the curves move upward as temperature increases; this is another factor which must be considered in d-c amplifiers and other designs where excessive shift of operating point cannot be tolerated. At high currents there is considerable crowding, denoting nonlinearity. It will also be noted that the saturation voltage is higher than for the germanium transistor.

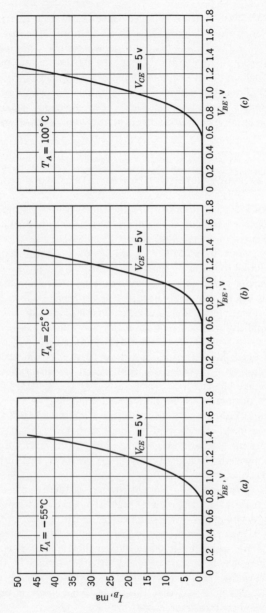

Fig. 5.5 Input characteristics, 2N1613 transistor.

CHARACTERISTIC CURVES 31

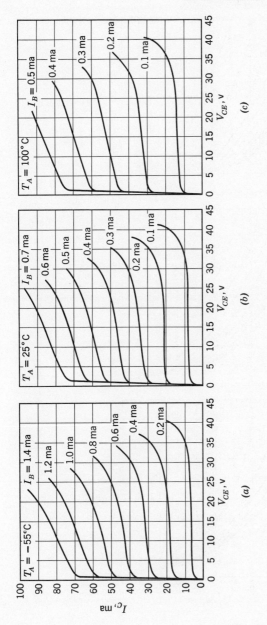

Fig. 5.6 Collector characteristics, 2N1613 transistor.

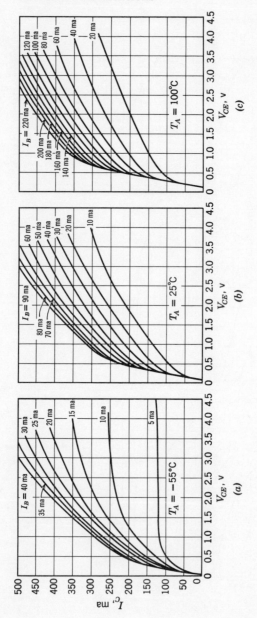

Fig. 5.7 Low-voltage, high-current characteristics, 2N1613 transistor.

5.5 Transistor Parameter Nomenclature

In earlier chapters we used the standard network parameter designations, z_{11}, y_{21}, h_{22}, etc. Industry has adopted an abbreviated set of parameters, coupled with added subscripts to denote the configuration. The relationship is as follows:

Subscript i denotes network subscript 11.
Subscript r denotes network subscript 12.
Subscript f denotes network subscript 21.
Subscript o denotes network subscript 22.

Subscripts b, e, c denote common base, common emitter, and common collector respectively.

Thus, z_{ib} means the same as z_{11_b}, etc. Since the h parameters are most widely used, we usually encounter the symbols h_{ib}, h_{ie}, h_{rb}, h_{re}, h_{fb}, h_{fe}, h_{ob}, h_{oe}, and their common-collector counterparts. Occasionally, however, we will encounter similar forms of the z and y parameters. In this text the abbreviated forms of parameter designations will be used wherever possible.

5.6 Relationship between Parameters and Characteristic Curves

The parameters we have discussed in earlier chapters are so-called small-signal parameters in transistor usage, implying that the a-c deviations around the operating points are relatively small, such that incremental techniques can be used to obtain the parameters from the characteristics. For example, the parameter h_{11} was defined as the input impedance with the output short-circuited. Obviously, we cannot supply the collector with the required d-c voltage if it is short-circuited to the base or emitter directly; however the effect of an a-c short-circuit is achieved if the d-c voltage between collector and base or emitter, as the case may be, is held constant. We can, therefore, define the input h parameter for the transistor as follows:

For the common-base configuration, h_{ib} is the incremental input impedance with the collector-base voltage held constant. For the common-emitter configuration, h_{ie} is the incremental input impedance with the collector-emitter voltage constant.

Using the above definitions, it becomes evident that the parameter h_{ib} would be the slope of the input curve for the common-base configuration e.g., Fig. 5.3, and that h_{ie} is the slope of the curves for the common-emitter configuration, e.g., Fig. 5.5. Since the slopes of these curves change with current, it is evident that the input h parameters do likewise, and it will be shown later that they are almost inversely proportional to emitter or base current respectively.

In similar manner, it can be deduced that the output h parameters are the negative slopes, or admittances, of the collector curves for constant input currents. The shape of these curves also indicates that the parameters h_{ob} and h_{oe} will be current- and voltage-dependent.

The feedback parameters h_{rb} and h_{re} may be found from the two families of curves, although a family of input curves is now required rather than one curve. It is the ratio of emitter-voltage change to collector-voltage change, maintaining constant emitter current or base current respectively. This parameter cannot be ascertained with any degree of accuracy from static characteristics.

Undoubtedly the most important parameter is the forward transfer parameter h_{fb} and its common-emitter counterpart h_{fe}. These can be obtained from the static characteristics, being the ratio of change in collector current to the change in emitter or base current, respectively, at a constant voltage level. For example, in Fig. 5.6b a change of base current from 0.3 to 0.4 ma at $V_{CE} = 15$ v produces a change of collector current from about 34.5 ma to about 46.5 ma, or a change of about 12 ma. The parameter h_{fe}, therefore, is 12/0.1 or 120.

One other parameter frequently encountered in transistor specifications is the so-called d-c beta, designated as h_{FE}. This is given by the direct ratio of the collector current at the point of interest to the base current producing it. In the example above, it would be $46.5/0.4 = 116$ at $I_B = 0.4$ ma, and $34.5/0.3 = 115$ at $I_B = 0.3$ ma. If the curves are reasonably uniform in spacing the difference between the small-signal value, h_{fe} and the d-c value h_{FE} will not be great, as was the case above. Under such circumstances, the d-c value can usually be used with little loss of accuracy.

5.7 Determination of Small-Signal Parameters

While the preceding section indicated how the small-signal parameters could be obtained from the static characteristics, it is also obvious that the resultant accuracy will leave much to be desired. In actual practice these parameters are measured by inserting small a-c signals into the appropriate circuits, then measuring the resultant a-c currents or voltages at the desired locations. For example, h_{fe} would be measured by applying an a-c voltage between base and emitter, measuring the a-c current going into the base, e.g., by means of a drop across a resistor in the base lead, then measuring the a-c current flowing in the collector circuit, again by the drop across a resistor. Since the collector is supposed to be a-c short-circuited for this measurement, the technique must introduce minimum impedance into the collector circuit. Similar techniques are employed to obtain the other small-signal parameters, by using small a-c voltages or currents applied to either the input or output terminals and measuring the resultant currents or voltages.

5.8 Specification Values of Transistor Parameters

In addition to the static characteristics, the various transistor manufacturers publish nominal, minimum, and maximum values of the various transistor parameters. Since, as pointed out above, these parameters are current- and voltage-dependent, the measurements must be made under some standardized conditions. The standard is an emitter current of 1 ma and collector-base voltage of 5 v, unless otherwise specified, room temperature (25°C), and a frequency of 270 cycles. Nominal values of the h parameters for the 2N43 and 2N1613 transistors, under these conditions are:

Parameter	2N43	2N1613	Units
h_{ib}	29	27	ohms
h_{ie}	—	2200	ohms
h_{rb}	5×10^{-4}	0.7×10^{-4}	
h_{re}	—	3.6×10^{-4}	
h_{fe}	42	55	
h_{ob}	0.8×10^{-6}	0.16×10^{-6}	mho
h_{oe}	—	12.5×10^{-6}	mho

In addition, the following d-c values are given:

For the 2N43:

At $\quad V_{CE} = -1$ v, $I_C = -20$ ma, $h_{FE} = 53$ design center
$\qquad -1$ v $\qquad -100$ ma $\qquad 48$

For the 2N1613:

At $\quad V_{CE} = 10$ v, $I_C = \quad 10$ ma, $h_{FE} = 80$
$\qquad\qquad\qquad\qquad\quad 150$ ma $\qquad 80$
$\qquad\qquad\qquad\qquad\quad 500$ ma $\qquad 55$

5.9 Variation of Parameters with Operating Point and Temperature

Since the parameters vary greatly with operating bias and also with temperature, some method must be employed to obtain their values at some point or temperature other than the standard specification values. This is accomplished by means of curves which show the variation of parameters with emitter current, collector voltage, and temperature. Figures 5.8, 5.9 and 5.10 show the variation of the parameters with emitter current, collector voltage, and temperature, respectively, for the 2N43 transistor. Figures 5.11, 5.12, and 5.13 show similar curves for the 2N1613. In addition, Fig. 5.14 shows the variation of the d-c parameter h_{FE} with collector current and temperature for the 2N1613. (Note that these curves are normalized to 5 ma and 10 v, not 1 ma, 5 v.)

As an example of how to use these curves, consider the 2N1613. Let us obtain the parameters at $I_E = 2$ ma, $V_{CB} = 2$ v. We can either use

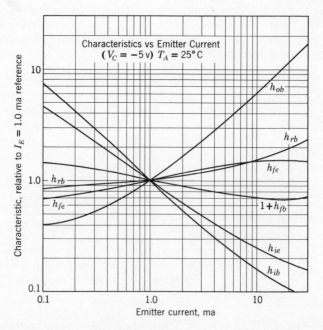

Fig. 5.8 Variation of parameters with emitter current, 2N43 transistor.

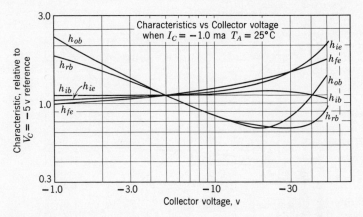

Fig. 5.9 Variation of parameters with collector voltage, 2N43 transistor.

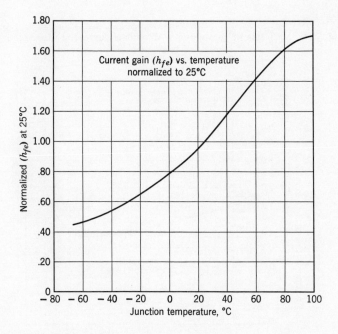

Fig. 5.10 Variation of h_{fe} with temperature, 2N43 transistor.

38   TRANSISTOR APPLICATIONS

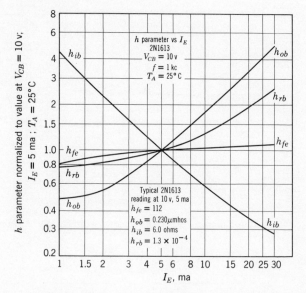

Fig. 5.11 Variation of parameters with emitter current, 2N1613 transistor.

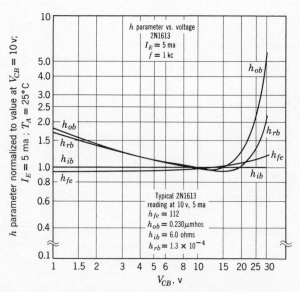

Fig. 5.12 Variation of parameters with collector voltage, 2N1613 transistor.

CHARACTERISTIC CURVES 39

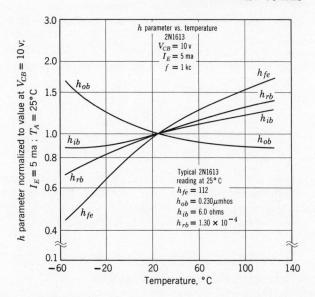

Fig. 5.13 Variation of parameters with temperature, 2N1613 transistor.

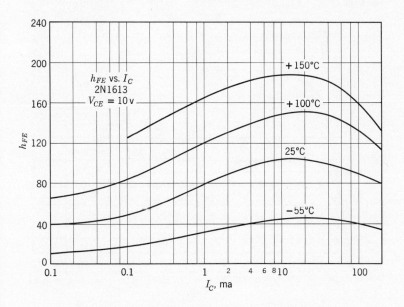

Fig. 5.14 h_{FE} versus collector current and temperature.

the 1-ma, 5-v values given above and correct them by the ratios of correction factors for the two sets of conditions, relative to 5 ma, 10 v, or we can use the values of parameters shown on Figs. 5.11–5.13 and make only the one set of corrections. Let us use the latter approach. Thus:

$$h_{fe} = 112 \times 0.92 \times 0.93 = 96$$

$$h_{ob} = 0.23 \times 0.55 \times 1.55 = 0.196 \ \mu\text{mhos}$$

$$h_{ib} = 6.0 \times 2.25 \times 1.0 = 13.5 \text{ ohms}$$

$$h_{rb} = 1.3 \times 10^{-4} \times 0.82 \times 1.50 = 1.6 \times 10^{-4}$$

In the above, the first correction factors are those for the change from 5 ma to 2, the second is for the change from 10 v to 2. If, in addition, a change in temperature from the standard 25° was indicated, a third correction, derived from Fig. 5.13, would have been required.

5.10 Static Characteristics for the Unipolar (Field-Effect) Transistor

This type of transistor has characteristics quite similar to those of the electron tube. Figure 5.15 shows the curves for a typical device. It

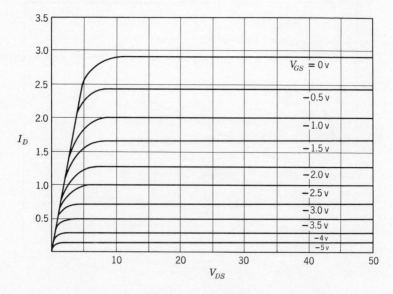

Fig. 5.15 Characteristics of field-effect transistor.

will be noted that a different terminology is used for these devices: The input electrode is called the gate, the other two are the source and drain, with the source roughly corresponding to the cathode of a tube or to the emitter of a common-emitter transistor stage, and the drain corresponding to the plate of the tube, collector of the transistor. The important parameters are the transconductance and output admittance. As evident from the figure, the latter is extremely low on the order of millimicromhos, hence its effect can usually be disregarded, and the device is considered to be one having only one parameter of importance, namely, the transconductance. This is the ratio of the change in drain current to the change of gate voltage producing it, for constant drain voltage. From Fig. 5.15 this can be calculated, and, for example, is 0.37 ma/0.5 v or 740 μmhos.

In matrix terminology, the unipolar transistor can best be represented by the y parameters, with all parameters approximately equal to zero, except y_{12}, which will equal the transconductance.

REFERENCES

1. Engineering Staff of Texas Instruments, Inc., *Transistor Circuit Design*, McGraw-Hill, New York, 1963, pp. 13–85.
2. Hunter, L. B., *Handbook of Semiconductor Electronics*, 2d Ed., McGraw-Hill, New York, 1962, pp. 11-1–11-23, 19-1–19-35.

6

Transistor Equivalent Circuits

6.1 The Ideal Transistor

The so-called ideal transistor is a purely hypothetical one in which the parasitic elements due to bulk resistances, leakage resistances, stray capacitance, etc., are absent, and for which the parameters are strictly functions of the base width, the minority carrier diffusion length, and the operating currents and voltages.

We introduce here certain new parameters of this ideal transistor:

r_ϵ = emitter diffusion resistance;
μ_0 = low-frequency reverse-voltage transfer ratio;
α_{b0} = low-frequency short-circuit forward current transfer ratio;
g_{cd} = collector diffusion conductance.

These will be recognized as simply new symbols for our h parameters, introduced here to indicate their special nature. If we use the designation (') for the h parameter of this ideal transistor, e.g., h_{ib}' and use a subscript 0 to indicate low-frequency values, we can show the following equivalences:

$$r_\epsilon = (h_{ib}')_0 \tag{6.1}$$

$$\mu_0 = (h_{rb}')_0 \tag{6.2}$$

$$\alpha_{b0} = -(h_{fb}')_0 \tag{6.3}$$

$$g_{cd} = (h_{ob}')_0 \tag{6.4}$$

Note that $(h_{fb}')_0$ is the *negative* of α_{b0}, due to an original convention

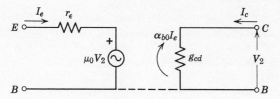

Fig. 6.1 Parameters of the ideal transistor.

showing the output-current generator in the opposite direction from the present practice.

Figure 6.1 shows these special parameters in the equivalent circuit corresponding to the h parameters. It should be remembered that this circuit has no actual existence and must be modified by the parasitic elements to represent an actual transistor. It should also be noted that the symbol I_e is used to designate the emitter current. The use of the lower case subscript here denotes a-c current, where the capital subscript used for currents and voltages previously denoted d-c values.

6.2 Relationship between the Ideal Parameters and the Transistor Construction

The ideal transistor parameters have the following approximate relationship to the transistor dimensions, diffusion length, and operating points:

$$r_\epsilon \simeq \frac{kT}{qI_E} \quad \text{(d-c emitter current here)} \tag{6.5}$$

$$\mu_0 \simeq \frac{p'}{(q/kT)\sinh p \cosh p} \tag{6.6}$$

$$\alpha_{b0} \simeq \frac{1}{\cosh p} \tag{6.7}$$

$$g_{cd} \simeq p'I_E \sinh p \tag{6.8}$$

where k = Boltzmanns constant,
 T = temperature, °K,
 q = electronic charge,
 $p = W/l$ (W = base width, l = minority carrier diffusion length),
 $p' = \partial p/\partial V_{CB'} = (\partial W/\partial V_{CB'})/l$.

It is interesting to note that the special parameters of the ideal transistor are interrelated, being hyperbolic functions mainly of the same physical details of the transistor. Analysis of the above equations reveals the following relationship:

$$\frac{\mu_0}{g_{cd}} \simeq \frac{\alpha_{b0} r_\epsilon}{2(1 - \alpha_{b0})} \tag{6.9}$$

6.3 The Parameters of the Real Transistor

To obtain these, we add certain parasitic elements to those of the ideal transistor:

r_b' = base-spreading resistance,
C_c = collector capacitance (the sum of the collector transition capacitance and the diffusion capacitance).

Furthermore, the collector diffusion conductance g_{cd} is modified by external leakage and both are lumped in a combined conductance g_c.

The equivalent circuit of the real transistor now is as shown in Fig. 6.2. Note that the output current generator is now shown as a negative current, to make its direction conform to the standard for network designations. The + sign on the input-voltage generator indicates the polarity when the collector is positive with respect to the base.

The circuit of Fig. 6.2 can now be analyzed by using the definitions of the h parameters to obtain the low-frequency parameters of the real transistor. For example, the input parameter h_{ib} for the common-base configuration is defined as the input impedance with the output short-circuited. Inspection of Fig. 6.2 reveals that, under this condition, the input-voltage generator vanishes, and the circuit reduces to the resistance r_ϵ in series with the parallel combination of r_b', g_c, C_c and the

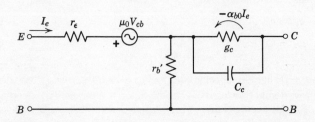

Fig. 6.2 The real transistor, including parasitic elements.

current generator $-\alpha_{b0}I_e$. Since the resistance of r_b' is many orders of magnitude lower than the reactance of C_c at low frequencies, and since g_{cd} will also be negligible, this circuit can be shown to reduce to the sum of r_ϵ and $r_b'(1 - \alpha_{b0})$. Thus, we obtain the following equations for the low-frequency h parameters of the real transistor, in terms of the elements of the ideal transistor plus the parasitic elements:

$$(h_{ib})_0 = r_\epsilon + r_b'(1 - \alpha_{b0}) \tag{6.10}$$

$$(h_{rb})_0 = \mu_0 + r_b'(g_c + j\omega C_c) \tag{6.11}$$

$$(h_{fb})_0 = -\alpha_{b0} \tag{6.12}$$

$$(h_{ob})_0 = g_c + j\omega C_c \tag{6.13}$$

(Normally at the specification frequency, e.g., 270 cps, the $j\omega C_c$ terms may be neglected.)

6.4 Derivation of the Parameters of the Ideal Transistor from Specification Sheet Values

We can use eqs. 6.10 through 6.13 to obtain the parameters r_ϵ, μ_0, α_{b0}, and g_c from the values of low-frequency h parameters given in the specification sheets. The parasitic base-spreading resistance r_b' can also be found to a reasonable approximation. The capacitance C_c is also usually given in the specifications. The transistor parameters are:

$$r_\epsilon = \frac{kT}{qI_E} \simeq \frac{25.6}{I_E} \text{ (at room temperature)} \qquad (I_E \text{ in ma}) \tag{6.14}$$

$$r_b' \simeq \frac{h_{ib} - r_\epsilon}{1 - \alpha_{b0}} \tag{6.15}$$

$$\alpha_{b0} = -(h_{fb})_0 \tag{6.16}$$

$$g_c \simeq h_{ob} \tag{6.17}$$

$$\mu_0 \simeq h_{rb} - r_b' g_c \tag{6.18}$$

One might wonder why go to all the trouble of obtaining special parameters when the specification sheets give the h parameters already. The answer, as will become evident later, lies in the necessity of obtaining values for the h parameters at high frequencies. Although some manufacturers supply values of the more important parameters as functions of frequency, this information is usually conspicuous by its absence. By obtaining the above special parameters from the low-frequency specification values, it becomes possible to compute reasonably good values of the h parameters at frequencies approaching the transistor cutoff frequency, and thereby calculate the performance of high-frequency amplifiers.

6.5 Common-Emitter Equivalent Circuit Using Special Parameters

Since the common-emitter configuration is the most widely used, it is convenient to have also an equivalent circuit for this configuration, using the parameters of the ideal transistor and the parasitic elements. Figure 6.3 shows such a circuit, consisting of three complex branches and one current generator. The elements of these branches and the current generator may be obtained from the following equations:

$$g_{b'e} \simeq \frac{1 - \alpha_{b0}}{r_\epsilon} \tag{6.19}$$

$$C_{b'e} \simeq \frac{1}{\omega_{ab} r_\epsilon} \tag{6.20}$$

$$g_{cb'} \simeq g_c \tag{6.21}$$

$$C_{cb'} \simeq C_c \tag{6.22}$$

$$g_{ec1} \simeq \frac{g_c}{1 - \alpha_{b0}} \tag{6.23}$$

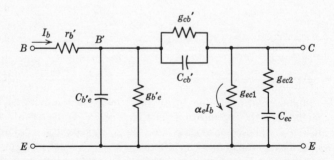

Fig. 6.3 Equivalent circuit for common-emitter configuration.

$$g_{ec2} \simeq C_c \omega_{ab} - \frac{g_c}{1 - \alpha_{b0}} \tag{6.24}$$

$$\alpha_e \simeq \frac{\alpha_{b0}}{1 - \alpha_{b0} + j\omega/\omega_{ab}} \tag{6.25}$$

In the above equations $\omega_{ab} = 2\pi$ times the frequency at which the magnitude of α ($-h_{fb}$) has decreased to 0.707 of its low-frequency value.

Since the circuit of Fig. 6.3 includes the capacitors corresponding to the various diffusion and transition capacitances of the transistor, it may be used with reasonable accuracy up to frequencies approaching the cutoff frequency ω_{ab}. We will have occasion to use this circuit and these parameters later when we consider high-frequency applications.

6.6 Determination of Network h Parameters from Ideal Transistor Parameters and Parasitic Elements

The previous section has indicated the method to be used to obtain the special parameters r_e, μ_0, α_{b0} and g_c from a knowledge of operating current and the low-frequency spec sheet h parameters. Having thus determined these special intrinsic parameters, it is useful to have a technique for reversing the process, deriving in this manner the h parameters for the actual transistor at any desired frequency. This method becomes necessary in designing high-frequency amplifiers, where the high-frequency values of the h parameters are not usually to be found in the specifications.

$$h_{ib} = \frac{r_e(1 + j\omega/2.4\omega_{ab})}{1 + j\omega/0.8\omega_{ab}} + r_b'\left[1 + \frac{(h_{fb})_0}{(1 + j\omega/\omega_{ab})(1 + j\omega/4\omega_{ab})}\right] \tag{6.26}$$

$$h_{rb} = \frac{\mu_0}{1 + j\omega/0.8\omega_{ab}} + r_b'[(h_{ob})_0 + j\omega C_c] \tag{6.27}$$

$$h_{fb} = \frac{(h_{fb})_0}{(1 + j\omega/\omega_{ab})(1 + j\omega/4\omega_{ab})} \tag{6.28}$$

$$h_{ob} = (h_{ob})_0 + j\omega C_c \tag{6.29}$$

The above equations include second-order factors intended to produce the proper phase relationships as well as magnitude. It is frequently adequate to use simpler expressions which are reasonably correct with respect to magnitude, although not entirely correct with respect to phase. These approximate equations are:

$$h_{ib} = \frac{1}{1 + j\omega/\omega_{ab}} \left[r_\epsilon + r_b' \left(1 - \alpha_{b0} + \frac{j\omega}{\omega_{ab}} \right) \right] \quad (6.30)$$

$$h_{rb} = (h_{rb})_0 + j\omega r_b' C_c \quad (6.31)$$

$$h_{fb} = \frac{(h_{fb})_0}{1 + j\omega/\omega_{ab}} \quad (6.32)$$

The equation for h_{ob} remains the same as in eq. 6.29.

As frequency increases still further, approaching the cutoff frequency, still further simplification becomes possible. At these frequencies:

$$h_{ib} \simeq \frac{r_\epsilon + (j\omega r_b')/\omega_{ab}}{1 + j\omega/\omega_{ab}} \quad (6.33)$$

$$h_{rb} \simeq j\omega r_b' C_c \quad (6.34)$$

$$h_{fb} \simeq -\frac{\alpha_{b0}}{1 + j\omega/\omega_{ab}} \quad (6.35)$$

$$h_{ob} \simeq j\omega C_c \quad (6.36)$$

The corresponding common-emitter parameters are:

$$h_{ie} \simeq r_b' - \frac{j\omega_{ab} r_\epsilon}{\omega} \quad (6.37)$$

$$h_{re} \simeq \omega_{ab} r_\epsilon C_c \quad (6.38)$$

$$h_{fe} \simeq \frac{j\omega_{ab}\alpha_{b0}}{\omega} \quad (6.39)$$

$$h_{oe} \simeq \omega_{ab} C_c \left(1 + \frac{j\omega}{\omega_{ab}} \right) \quad (6.40)$$

6.7 Converting Common-Base Parameters to Common-Emitter and Common-Collector Parameters

Tables 6.1 and 6.2 permit conversion from one set of transistor parameters to any of the others. Table 6.1 gives more accurate results than Table 6.2; however, the accuracy achieved with Table 6.2 is more than sufficient for the great majority of cases.

Table 6.1 Matrix Interrelations of the Transistor h Parameters

	In Terms of Common-Base Parameters	In Terms of Common-Emitter Parameters	In Terms of Common-Collector Parameters
$[h_b]$	—	$\dfrac{1}{1 + h_{fe} + \Delta^h_e - h_{re}} \begin{bmatrix} h_{ie}; & (\Delta^h_e - h_{re}) \\ -(\Delta^h_e + h_{fe}); & h_{oe} \end{bmatrix}$ $\Delta^h_b = \Delta^h_e/(1 + h_{fe} + \Delta^h_e - h_{re})$	$\dfrac{1}{\Delta^h_c} \begin{bmatrix} h_{ic}; & (\Delta^h_c + h_{fc}) \\ -(\Delta^h_c - h_{rc}); & h_{oc} \end{bmatrix}$ $\Delta^h_b = (1 + h_{fc} + \Delta^h_c - h_{rc})/\Delta^h_c$
$[h_e]$	$\dfrac{1}{1 + h_{fb} + \Delta^h_b - h_{rb}} \begin{bmatrix} h_{ib}; & (\Delta^h_b - h_{rb}) \\ -(h_{fb} + \Delta^h_b); & h_{ob} \end{bmatrix}$ $\Delta^h_e = \Delta^h_b/(1 + h_{fb} + \Delta^h_b - h_{rb})$	—	$\begin{bmatrix} h_{ic}; & (1 - h_{rc}) \\ -(1 + h_{fc}); & h_{oc} \end{bmatrix}$ $\Delta^h_e = 1 + h_{fc} + \Delta^h_c - h_{rc}$
$[h_c]$	$\dfrac{1}{1 + h_{fb} + \Delta^h_b - h_{rb}} \begin{bmatrix} h_{ib}; (1 + h_{fb}) \\ -(1 - h_{rb}); h_{ob} \end{bmatrix}$ $\Delta^h_c = 1/(1 + h_{fb} + \Delta^h_b - h_{rb})$	$\begin{bmatrix} h_{ie}; & (1 - h_{re}) \\ -(1 + h_{fe}); & h_{oe} \end{bmatrix}$ $\Delta^h_c = 1 + h_{fe} + \Delta^h_e - h_{re}$	—

Table 6.2 Matrix Interrelations of Transistor h Parameters (Approximate)

	In Terms of Common-Base Parameters	In Terms of Common-Emitter Parameters	In Terms of Common-Collector Parameters
$[h_b]$	—	$\dfrac{1}{1+h_{fe}}\begin{bmatrix} h_{ie};\ (\Delta^h_e - h_{re}) \\ -h_{fe};\ h_{oe} \end{bmatrix}$ $\Delta^h_b \simeq \Delta^h_e/(1+h_{fe})$	$\dfrac{-1}{h_{fc}}\begin{bmatrix} h_{ic};\ (\Delta^h_c + h_{fc}) \\ h_{rc}h_{fc};\ h_{oc} \end{bmatrix}$ $\Delta^h_b \simeq -(h_{fc} + \Delta^h_c)/h_{fc}$
$[h_e]$	$\dfrac{1}{1+h_{fb}}\begin{bmatrix} h_{ib};\ (\Delta^h_b - h_{rb}) \\ -h_{fb};\ h_{ob} \end{bmatrix}$ $\Delta^h_e = \Delta^h_b/(1+h_{fb})$	—	$\begin{bmatrix} h_{ic};\ (1-h_{rc}) \\ -h_{fc};\ h_{oc} \end{bmatrix}$ $\Delta^h_e = h_{fc} + \Delta^h_c$
$[h_c]$	$\dfrac{1}{1+h_{fb}}\begin{bmatrix} h_{ib};\ (1+h_{fb}) \\ -1;\ h_{ob} \end{bmatrix}$ $\Delta^h_c = 1/(1+h_{fb})$	$\begin{bmatrix} h_{ie};\ 1 \\ -(1+h_{fe});\ h_{oe} \end{bmatrix}$ $\Delta^h_c = 1 + h_{fe}$	—

As an example of the use of these tables, let us obtain the equation for the parameter h_{ie}. From Table 6.1:

$$h_{ie} = \frac{h_{ib}}{1 + h_{fb} + \Delta^h_b - h_{rb}} \tag{6.41}$$

Substituting $h_{ib}h_{ob} - h_{rb}h_{fb}$ for Δ^h_b and solving, we obtain:

$$h_{ie} = \frac{h_{ib}}{(1+h_{fb})(1-h_{rb}) + h_{ib}h_{ob}} \tag{6.42}$$

From Table 6.2 the approximate equation is

$$h_{ie} \simeq \frac{h_{ib}}{1 + h_{fb}} \tag{6.43}$$

As an indication of the degree of approximation, take the specification values for the 2N43:

$h_{ib} = 29$ ohms $h_{rb} = 5 \times 10^{-4}$ $h_{ob} = 0.8 \times 10^{-6}$ mho

h_{fb} is obtained from h_{fe} by the relation $h_{fb} = -h_{fe}/(1+h_{fe})$ and is -0.9768.

$$\Delta^h_b = 29 \times 0.8 \times 10^{-6} + 0.9768 \times 5 \times 10^{-4} = 5.116 \times 10^{-4}$$

The exact equation gives

$$h_{ie} = 29 \div (0.0232 \times 0.9995 + 0.0005116) = 1223 \text{ ohms}$$

The approximate equation gives

$$h_{ie} = 29 \div 0.0232 = 1250 \text{ ohms}$$

Thus, the approximate equation gives an answer within 2%, usually more than sufficiently accurate. One occasion when this will not necessarily be true is if external impedance is inserted in series with the emitter lead. This materially changes the value of the determinant and makes it necessary to use the more exact equations.

REFERENCE

1. Shea, R. F. et al., *Transistor Circuit Engineering*, Wiley, New York 1957, pp. 21–49.

7

Bias

7.1 Collector Cutoff Current

The collector cutoff current I_{CBO} plays a considerable part in the determination of the operating point of a transistor amplifier, and thus in the constancy of impedances, amplification, and gain with temperature. This current I_{CBO} is the collector-to-base current flowing, with the emitter open. Figures 7.1 and 7.2 show the variation of this current with temperature for the 2N43 and 2N1613 respectively. It will be noted that this variation is essentially logarithmic, indicating that the cutoff current increases exponentially with temperature. There is an approximately 10:1 increase in I_{CBO} for an increase of 30° for the 2N43 germanium transistor, and about the same change in 45° for the 2N1613, although at a higher collector voltage. At the lower voltage at which the first curve was taken, the rate of change would be comparable. This does not mean that the values of I_{CBO} are comparable, however, since the rated value for the silicon transistor is much lower than that of the germanium.

In estimating the value of I_{CBO} some approximation is necessary in converting from the spec values to the values corresponding to actual design. For example, the rating for the 2N43 is a design center value of -8 μa at 45 v, maximum -16 μa. For the 2N1613, the corresponding values are a maximum value of 10 mμa at 60 v. Thus it is seen that the silicon transistor has a value three orders of magnitude lower than that of the germanium transistor. As noted above, an estimate is necessary to obtain the value at some lower voltage, say, at 5 v. Since the current I_{CBO} is composed of both leakage current and the expo-

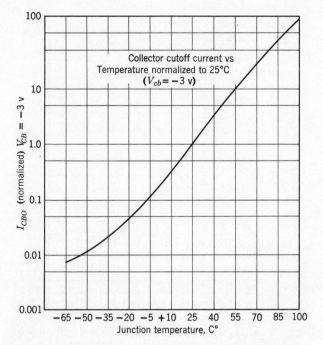

Fig. 7.1 Variation of I_{CBO} with temperature, 2N43 transistor.

nentially varying saturation current, it will be neither linearly proportional to collector voltage nor follow an exponential relationship. A reasonable estimate is to assume half of the current is leakage, hence proportional to voltage, and the remainder varying exponentially. By using this rule, we would have values of I_{CBO} of about -3 μa for the 2N43 and 4 mμa for the 2N1613 at a collector voltage of 5 v.

7.2 Effect of Variation of V_{BE}

Figure 5.5 showed the variation of V_{BE} with temperature for the 2N1613, and indicated a shift of about 1.3 mv/°C. It will be found that both silicon and germanium transistors behave alike in this respect, with shifts between 1.0 and 2.0 mv/°C. This shift must be considered in the design of d-c amplifiers and in other designs where small shifts of operating point cannot be tolerated.

54   TRANSISTOR APPLICATIONS

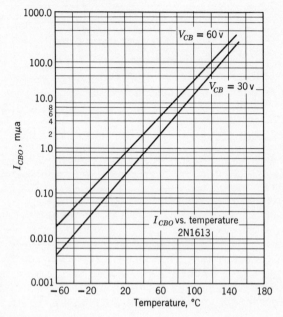

Fig. 7.2   Variation of I_{CBO} with temperature, 2N1613 transistor.

7.3   Methods of Biasing Transistors

Two techniques are generally used to provide the necessary biases for the transistor—one employs separate batteries for the emitter and collector supplies, the other obtains both biases from the same battery. These arrangements are referred to as two-battery and single-battery bias respectively.   Figures 7.3 through 7.6 illustrate these arrangements. In Fig. 7.3 a battery V_{EE} supplies the emitter through a resistor R_1, and the collector is fed from the battery V_{CC} through R_C.   This arrangement might be encountered in the common-base configuration, where R_1 might be the d-c resistance of a transformer, for example.   Figure 7.4 shows a variation where an additional resistor R_2 is inserted in series with the base.   This might be the case for the common-emitter configuration, where R_2 might be the transformer resistance.

Figures 7.5 and 7.6 show how the biases are obtained from the one battery by using a voltage divider consisting of resistors R_2 and R_3 to apply a voltage to the base.   Again, the two figures correspond to the

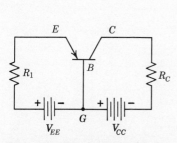
Fig. 7.3 Two-battery biasing, common-base configuration.

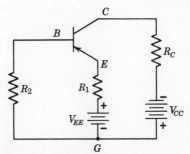
Fig. 7.4 Two-battery biasing, common-emitter configuration.

common-base and common-emitter configurations respectively, although the d-c connections are identical. It will also be noted that the voltage across R_2 subtracts from V_{CC}, so that the collector-base voltage is less by that amount.

No by-pass capacitors are shown on these figures, since they are intended to only illustrate d-c biasing methods. In actual practice, by-passes would be employed to avoid a-c degeneration or attenuation.

7.4 Equations for the Two-Battery Arrangement

The following equations are based on the assumption that the resistance inserted in series with the emitter is sufficient to swamp the effect

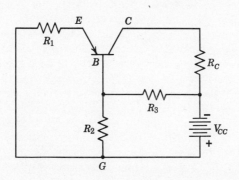
Fig. 7.5 Single-battery biasing, common-base configuration.

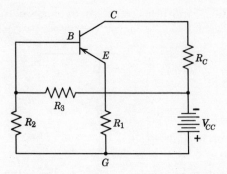

Fig. 7.6 Single-battery biasing, common-emitter configuration.

of V_{BE}. In a later portion of this chapter, a graphical method will be presented which takes this factor into account.

By using the symbols of Fig. 7.4, the currents and voltages may be calculated as follows (the circuit of Fig. 7.3 is a special case of Fig. 7.4, where $R_2 = 0$):

$$I_E = \frac{V_{EE} - R_2 I_{CBO}}{R_1 + R_2(1 + h_{fb})} \tag{7.1}$$

$$I_B = -\left[\frac{R_1 I_{CBO} + V_{EE}(1 + h_{fb})}{R_1 + R_2(1 + h_{fb})}\right] \tag{7.2}$$

$$I_C = \frac{(R_1 + R_2) I_{CBO} + h_{fb} V_{EE}}{R_1 + R_2(1 + h_{fb})} \tag{7.3}$$

$$V_{BG} = V_{EG} = R_2 \left[\frac{R_1 I_{CBO} + V_{EE}(1 + h_{fb})}{R_1 + R_2(1 + h_{fb})}\right] \tag{7.4}$$

$$V_{CG} = V_{CC} - R_C \left[\frac{(R_1 + R_2) I_{CBO} + h_{fb} V_{EE}}{R_1 + R_2(1 + h_{fb})}\right] \tag{7.5}$$

$$V_{CB} = V_{CC} - V_{EE} \left[\frac{R_2(1 + h_{fb}) + h_{fb} R_C}{R_1 + R_2(1 + h_{fb})}\right]$$
$$- I_{CBO} \left[\frac{R_1 R_2 + R_C(R_1 + R_2)}{R_1 + R_2(1 + h_{fb})}\right] \tag{7.6}$$

In addition, there are several stability factors which provide a measure of the variability of the operating point with temperature and with supply voltages. These factors give the variations in terms of I_{CBO} the variation of which with temperature can be obtained from curves such as shown in Figs. 7.1 and 7.2, and of the supply voltages V_{CC} and V_{EE}.

$$S_{IE} = \frac{dI_E}{dI_{CBO}} = \frac{-R_2}{R_1 + R_2(1 + h_{fb})} \tag{7.7}$$

$$S_{IC} = \frac{dI_C}{dI_{CBO}} = \frac{R_1 + R_2}{R_1 + R_2(1 + h_{fb})} \tag{7.8}$$

$$S_V = \frac{dV_{CB}}{dI_{CBO}} = -\left[\frac{R_1R_2 + R_C(R_1 + R_2)}{R_1 + R_2(1 + h_{fb})}\right] \tag{7.9}$$

$$S_{V_1} = \frac{dV_{CB}}{dV_{CC}} = 1 \tag{7.10}$$

$$S_{V_2} = \frac{dV_{CB}}{dV_{EE}} = -\frac{R_C h_{fb} - R_2(1 + h_{fb})}{R_1 + R_2(1 + h_{fb})} \tag{7.11}$$

Note: V_{CC} will be negative, V_{EE} positive, and I_{CBO} negative in *p-n-p* transistors. The reverse will be true with *n-p-n* transistors.

In the above equations, G refers to a virtual ground at the connection between the low side of the resistor R_2 and the common side of the two batteries.

7.5 Equations for the Single-Battery Arrangement

Referring to Figs. 7.5 and 7.6 the currents, voltages, and stability factors may be obtained from the following equations:

$$I_E = \frac{-R_2 V_{CC}}{R_1R_2 + R_1R_3 + R_2R_3(1 + h_{fb})}$$
$$\qquad - \frac{R_2R_3 I_{CBO}}{R_1R_2 + R_1R_3 + R_2R_3(1 + h_{fb})} \tag{7.12}$$

$$I_B = \frac{R_2 V_{CC}(1 + h_{fb})}{R_1R_2 + R_1R_3 + R_2R_3(1 + h_{fb})}$$
$$\qquad - \frac{R_1(R_2 + R_3) I_{CBO}}{R_1R_2 + R_1R_3 + R_2R_3(1 + h_{fb})} \tag{7.13}$$

$$I_C = \frac{-h_{fb} R_2 V_{CC}}{R_1R_2 + R_1R_3 + R_2R_3(1 + h_{fb})}$$
$$\qquad + \frac{(R_1R_2 + R_1R_3 + R_2R_3) I_{CBO}}{R_1R_2 + R_1R_3 + R_2R_3(1 + h_{fb})} \tag{7.14}$$

58 TRANSISTOR APPLICATIONS

$$V_{BG} = V_{EG} = \frac{R_1R_2V_{CC}}{R_1R_2 + R_1R_3 + R_2R_3(1 + h_{fb})}$$
$$+ \frac{R_1R_2R_3I_{CBO}}{R_1R_2 + R_1R_3 + R_2R_3(1 + h_{fb})} \quad (7.15)$$

$$V_{CG} = V_{CC}\left[1 + \frac{R_2R_Ch_{fb}}{R_1R_2 + R_1R_3 + R_2R_3(1 + h_{fb})}\right]$$
$$- \frac{R_C(R_1R_2 + R_1R_3 + R_2R_3)I_{CBO}}{R_1R_2 + R_1R_3 + R_2R_3(1 + h_{fb})} \quad (7.16)$$

$$V_{CB} = V_{CC}\left[\frac{R_1R_3 + R_2R_3(1 + h_{fb}) + R_2R_Ch_{fb}}{R_1R_2 + R_1R_3 + R_2R_3(1 + h_{fb})}\right]$$
$$- I_{CBO}\left[\frac{R_1R_2R_3 + R_C(R_1R_2 + R_1R_3 + R_2R_3)}{R_1R_2 + R_1R_3 + R_2R_3(1 + h_{fb})}\right] \quad (7.17)$$

$$S_{I_E} = \frac{dI_E}{dI_{CBO}} = \frac{-R_2R_3}{R_1R_2 + R_1R_3 + R_2R_3(1 + h_{fb})} \quad (7.18)$$

$$S_{I_C} = \frac{dI_C}{dI_{CBO}} = \frac{R_1R_2 + R_1R_3 + R_2R_3}{R_1R_2 + R_1R_3 + R_2R_3(1 + h_{fb})} \quad (7.19)$$

$$S_V = \frac{dV_{CB}}{dI_{CBO}} = -\left[\frac{R_1R_2R_3 + R_C(R_1R_2 + R_1R_3 + R_2R_3)}{R_1R_2 + R_1R_3 + R_2R_3(1 + h_{fb})}\right] \quad (7.20)$$

$$S_{V_1} = \frac{dV_{CB}}{dV_{CC}} = \frac{R_1R_3 + R_2R_3(1 + h_{fb}) + R_2R_Ch_{fb}}{R_1R_2 + R_1R_3 + R_2R_3(1 + h_{fb})} \quad (7.21)$$

Note: the same polarity conventions apply as for the two-battery arrangement.

7.6 Graphical Determination of Operating Point

A more accurate determination of the operating point, including the effect of the actual value of the emitter-base voltage V_{EB}, may be obtained by using the static characteristics of the transistor. Figure 7.7 shows the single-battery circuit once more, with the various currents and voltages identified. The steps to follow in obtaining the operating point are as follows:

1. Construct a family of I_C versus V_{CE} curves for the region of interest from the manufacturer's specification sheet.

BIAS 59

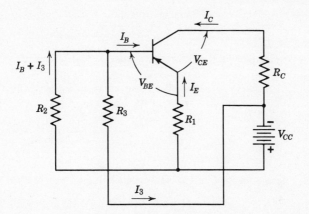

Fig. 7.7 Common-emitter stage, showing voltages and currents.

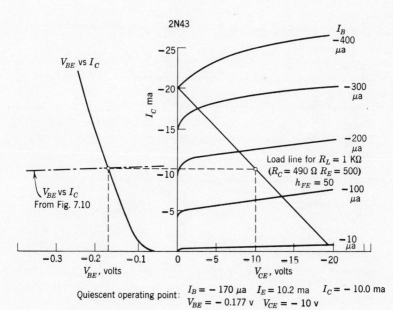

Quiescent operating point: $I_B = -170\ \mu a$ $I_E = 10.2\ ma$ $I_C = -10.0\ ma$
$V_{BE} = -0.177\ v$ $V_{CE} = -10\ v$

Fig. 7.8 Input and output characteristics, with load line.

2. Draw the effective load line from the point V_{CC}, having a slope equal to $R_C + R_1(h_{FE} + 1)/h_{FE}$. (*Note:* h_{FE} is the d-c value, equal to the ratio I_C/I_B.)
3. Construct a plot of I_C and V_{CE} versus I_B from points lying on this load line.
4. From these curves and the spec sheet curve of V_{BE} versus I_B plot V_{BE} versus I_C for the points on the load line.
5. Also plot V_{BE} versus I_C as obtained from the following equation:

$$V_{BE} = \frac{R_2 V_{CC}}{R_2 + R_3} - \frac{I_C}{h_{FE}} R_1(h_{FE} + 1) + \frac{R_2 R_3}{R_2 + R_3} \quad (7.22)$$

The intersection of the last two curves will determine the operating point.

The above procedure is illustrated in Figs. 7.8, 7.9, and 7.10. Figure 7.8 shows the collector family for the 2N43 transistor, together with a load line corresponding to the following conditions:

$V_{CC} = -20$ v $\quad\quad h_{FE} = 50$

$R_1 = 500$ ohms $\quad\quad R_2 = 10,000$ ohms

$R_3 = 20,000$ ohms $\quad\quad R_C = 490$ ohms

These values give an effective load line resistance of 1000 ohms, according to step 2 above.

Figure 7.9 shows the variation of I_C and V_{CE} versus I_B, obtained by picking various points off the load line. Knowing the values of V_{CE} corresponding to each value of I_B and also the corresponding values of

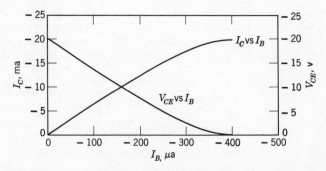

Fig. 7.9 Collector current and collector-emitter voltage versus base current.

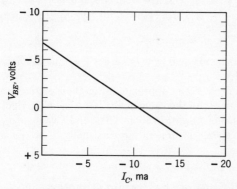

Fig. 7.10 Calculated V_{BE} versus I_C for input circuit.

V_{BE} from the emitter family, the curve of V_{BE} versus I_C was plotted in the left-hand quadrant of Fig. 7.8.

Figure 7.10 is a plot of eq. 7.22, and a small segment of this curve is also plotted as the dot-dash line in the left-hand quadrant of Fig. 7.8. The intersection of these two curves gives the operating point, which turns out to be:

$$I_C = -10 \text{ ma} \qquad I_E = 10.2 \text{ ma} \qquad I_B = -0.17 \text{ ma}$$
$$V_{CE} = -10 \text{ v} \qquad V_{BE} = -0.177 \text{ v}$$

Usually the approximate method described in eqs. 7.1 through 7.21 will supply reasonably accurate values for the majority of applications. However, if greater accuracy is required, the graphical method will usually supply it. As an example of the accuracies to be expected from the foregoing two methods of obtaining bias currents and voltages, let us calculate I_C, I_B, and V_{CE} from the equations in Section 7.5 for the above conditions. Substituting the given values, we obtain:

$$I_C = -0.98 \times 10{,}000 \times 20 \div 19 \times 10^6 - 215 \times 10^6 \times 3 \times 10^{-3}$$
$$\div 19 \times 10^6$$
$$= (-10.32 - 0.03)10^{-3} = -10.35 \text{ ma}$$
$$I_B = 10{,}000(-20)0.02 \div 19 \times 10^6 + 500 \times 30{,}000 \times 3 \times 10^{-6}$$
$$\div 19 \times 10^6$$
$$= -0.210 \times 10^{-3} + 0.002 \times 10^{-3} = -0.208 \text{ ma}$$

$$V_{CE} = V_{CB} = -20(14 \times 10^6 - 0.98 \times 4.9 \times 10^6) \div 19 \times 10^6$$
$$+ 3 \times 10^{-6}(100{,}000 + 490 \times 215 \times 10^6) \div 19 \times 10^6$$
$$= -9.67 + 0.02 = -9.65 \text{ v}$$

(*Note:* V_{CE} assumed equal to V_{CB} since V_{EB} is neglected in this approximate method.)

We see, thus, that the approximate method produces a value of I_C within less than 4% of the value obtained by the more rigorous graphical method, a value of I_B within about 20%, and a value of V_{CE} within less than 4%. These accuracies are normally good enough for most purposes.

7.7 Significance of the Stability Factors

The various stability factors given above provide a means of determining the shift of operating point with temperature, and thus the variation of gain or impedance. Although the various transistor parameters themselves are not excessively responsive to temperature, they do vary considerably with emitter current and collector-emitter voltage. If these vary excessively with temperature, they will, in turn, produce a variation of the transistor parameters. It is this secondary effect of temperature which contributes to the apparent temperature sensitivity of transistor circuits. Thus, if we have criteria by which we can anticipate the effects of variation of I_{CBO} with temperature, we can then determine how much shift in this quantity can be tolerated. The stability factors give us the relationships between variation of I_{CBO} and, for example, I_C. Knowing how much variation we can tolerate in I_C before we either produce excessive effect on such circuit parameters as input impedance or gain, or produce excessive distortion in power stages, and knowing the variation of I_{CBO} with temperature, we can then either set limits for temperature or ascertain the necessary bias arrangement and power loss required to permit operation over a stipulated temperature range.

The lower the stability factors, the less will be the variations of the operating currents and voltages with temperature and with supply voltages. However, we do not obtain this for nothing, and lower stability factors are only obtained by stiffening bias supplies, which means increased power consumption in the bias network. Thus, there is always a compromise between stability and efficiency. In low-power stages this is not serious; however, in power stages it becomes an important factor, and it is usually more economical to obtain freedom from temperature drift by means of such devices as thermistors and diodes, as will be shown in the chapters on power stages.

In the example above, the stability factor S_{IE} calculates to be 10.5. The significance of this is that the emitter current will change by about ten times as much as the change in I_{CBO}. Thus, if we can only tolerate a drift of, say, 0.5 ma or 5% in I_E, we are restricted to a drift of about 50 μa in I_{CBO}. Using the value of 3 μa for the 2N43 at 25°C, this implies a change of about 17:1. Referring to Fig. 7.1 shows that this change occurs for an increase to about 57°C. If we require better stability than this, we must increase the power dissipation in the bias network or, alternatively, use silicon transistors with their lower I_{CBO}.

REFERENCES

1. Hunter, L. B., *Handbook of Semiconductor Electronics*, 2nd Ed., McGraw-Hill, New York 1964, pp. 11-67–11-82.
2. Shea, R. F. (Ed.), *Transistor Circuit Engineering*, New York, Wiley, 1957, pp. 52–71.

8
Characteristics of the Single-Stage Amplifier

8.1 Introduction

In this chapter we will tie together much of the preceding material and give the applicable equations for the single-transistor stage in its three configurations, so that it will then be unnecessary to make any transformations from the specification parameters before being able to use them to determine circuit performance.

Figures 4.4, 4.5, and 4.6 showed the equivalent circuits for the three most common network arrangements—the z, y, and h circuits. Of these the h is the most commonly encountered, and will be used almost exclusively in the remainder of this book. Whenever it becomes necessary to use the other forms, their values may be obtained from the h

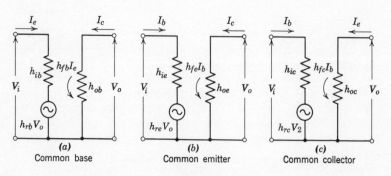

Fig. 8.1

CHARACTERISTICS OF THE SINGLE-STAGE AMPLIFIER 65

parameters by means of the transformation relationships given in Tables 1.1 and 1.2 of Chapter 1.

Although the equivalent circuits for the three configurations have all the same form, their constituent elements differ. Figures 8.1a, b, and c, show these three circuits. In the common-base circuit, the parameters bear the b subscripts, following the letter subscripts indicating the respective h parameters (see Section 5.5 for identification of parameter symbols). The input current is I_e (the lower-case subscript now indicates an a-c or small-signal quantity), the output current is I_c. In the common-emitter circuit, the parameter subscripts e are used, the input current is I_b, the output current I_c. In the emitter-follower or common-collector stage, the c subscript is used for the parameters, I_b is the input, I_e the output.

8.2 The Equivalent Circuits in Terms of Specification Parameters

Most specification sheets do not give all forms of h parameters, the most frequently given being the common-base parameters h_{ib}, h_{rb}, and h_{ob} and the common-emitter parameter h_{fe}. Occasionally the common-emitter input parameter h_{ie} is also given. The missing parameters can be obtained by means of the matrix interrelationships given in Tables 6.1 and 6.2. In most cases the approximate equations of Table 6.2 are sufficiently accurate and will be used in the following chapters unless otherwise noted.

Since all but one of the common-base parameters are obtainable from the specification sheet, we need only obtain the parameter h_{fb} from the spec value h_{fe}. From Table 6.2 we obtain

$$h_{fb} = -\frac{h_{fe}}{1 + h_{fe}} \tag{8.1}$$

We can thus construct a revised equivalent circuit for the common-base configuration by using those parameters to be found in the specs. This is shown in Fig. 8.2. The only change from Fig. 8.1a is the substitution of the current generator using the h_{fe} parameter.

In a similar manner we can obtain an equivalent circuit for the common-emitter configuration, in terms of the available parameters. This is shown in Fig. 8.3. In obtaining this circuit, two manipulations were employed:

Since $h_{fb} = -h_{fe}/(1 + h_{fe})$, $1 + h_{fb} = 1/(1 + h_{fe})$ and, therefore, $1 + h_{fe}$ can be substituted for $1/(1 + h_{fb})$.

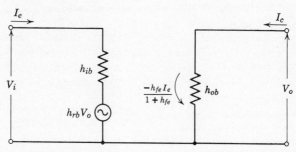

Fig. 8.2 Common-base circuit using common-emitter current generator.

To obtain the h_r parameter, we expand the interrelationship given in Table 6.2:

$$\frac{\Delta^h{}_b - h_{rb}}{1 + h_{fb}} = (h_{ib}h_{ob} - h_{rb}h_{fb} - h_{rb})(1 + h_{fe})$$

$$= h_{ib}h_{ob}(1 + h_{fe}) - h_{rb}(1 + h_{fb})(1 + h_{fe})$$

$$= h_{ib}h_{ob}(1 + h_{fe}) - h_{rb}$$

Thus, in the common-emitter configuration, the input short-circuit impedance and the output open-circuit admittance are both increased by the factor $1 + h_{fe}$, while the feedback voltage generator is a function of all four parameters.

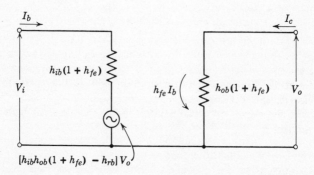

Fig. 8.3 Common-emitter equivalent circuit, using common-base and common-emitter parameters.

CHARACTERISTICS OF THE SINGLE-STAGE AMPLIFIER 67

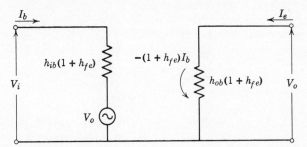

Fig. 8.4 Common-collector circuit, using common-base and common-emitter parameters.

Figure 8.4 shows the equivalent circuit for the common-collector configuration. The input and output parameters are the same as for the common-emitter configuration; however, the current generator is greater by one than the parameter h_{fe} and the phase is reversed. The voltage generator is simply the output voltage V_o.

8.3 Limitations on Use of Equivalent Circuits

The equivalent circuits shown in Figs. 8.2, 8.3, and 8.4 are general in nature in that they can be used under any conditions where the parameters are known. It should be borne in mind, however, that the spec sheet values are for low frequencies and, consequently, cannot be used much above the audio range. To determine the values at higher frequencies, the special parameters described in Chapter 6 must be obtained unless the manufacturers spec sheet gives high-frequency parameters. The high-frequency values of the desired h parameters are then obtained by using the conversion equations given in Chapter 6, and the performance of the stage can then be determined in the same manner as for lower frequencies, except that the parameters will usually be complex. This subject will be covered in greater detail in the chapters on high-frequency amplifiers.

8.4 Properties of Terminated Common-Base Stage

By using the equations given in Table 3.1 and the parameter values shown in Fig. 8.2, we can determine the properties of this stage. For

68 TRANSISTOR APPLICATIONS

example, the input impedance is given as $(h_{11} + \Delta^h Z_l)/(1 + h_{22}Z_l)$. Substituting h_{ib} for h_{11}, $h_{ib}h_{ob} - h_{rb}h_{fb}$ for Δ^h and h_{ob} for h_{22} and further substituting $-h_{fe}/(1 + h_{fe})$ for h_{fb}, we obtain the following equation, in terms of specification parameters:

$$Z_i = \frac{h_{ib} + [h_{ib}h_{ob} + h_{rb}h_{fe}/(1 + h_{fe})]Z_l}{1 + h_{ob}Z_l} \qquad (8.2)$$

Table 8.1 gives the properties of the terminated common-base stage, determined in the above manner. In this table, A_i is the current amplification, A_v the voltage amplification, G the gain through the stage

Table 8.1   Properties of Terminated Common-Base Stage

$$Z_i = \frac{h_{ib} + [h_{ib}h_{ob} + h_{rb}h_{fe}/(1 + h_{fe})]Z_l}{1 + h_{ob}Z_l}$$

$$Z_o = \frac{h_{ib} + Z_g}{h_{ob}Z_g + h_{ib}h_{ob} + h_{rb}h_{fe}/(1 + h_{fe})}$$

$$A_i = \frac{-h_{fe}}{(1 + h_{fe})(1 + h_{ob}Z_l)}$$

$$A_v = \frac{h_{fe}Z_l}{h_{ib}(1 + h_{fe}) + (1 + h_{fe})[h_{ib}h_{ob} + h_{rb}h_{fe}/(1 + h_{fe})]Z_l}$$

$$G = \frac{(h_{fe})^2 R_l}{(1 + h_{fe})^2 \{h_{ib} + [h_{ib}h_{ob} + h_{rb}h_{fe}/(1 + h_{fe})]R_l\}(1 + h_{ob}R_l)}$$

$$G_t = \frac{4(h_{fe})^2 R_g}{R_l\{(1 + h_{fe})[R_g(h_{ob} + 1/R_l) + h_{ib}/R_l + h_{ib}h_{ob} + h_{rb}h_{fe}/(1 + h_{fe})]\}^2}$$

$$G_m = \frac{(h_{fe})^2}{\{(1 + h_{fe})[\sqrt{h_{ib}h_{ob} + h_{rb}h_{fe}/(1 + h_{fe})} + \sqrt{h_{ib}h_{ob}}]\}^2}$$

$$R_{lm} = \sqrt{\frac{h_{ib}}{h_{ob}[h_{ib}h_{ob} + h_{rb}h_{fe}/(1 + h_{fe})]}}$$

$$R_{im} = \sqrt{\frac{h_{ib}[h_{ib}h_{ob} + h_{rb}h_{fe}/(1 + h_{fe})]}{h_{ob}}}$$

alone, G_t the transducer gain when operating from a source having a resistance R_g, G_m is the maximum possible gain, when both input and output are matched, R_{lm} is the matched value of load for this condition, and R_{im} the matched source impedance.

8.5 Properties of Terminated Common-Emitter and Common-Collector Stages

The equations for the common-emitter and common-collector stages are presented in Tables 8.2 and 8.3 respectively, and also in terms of the specification sheet parameters.

Table 8.2   Properties of the Terminated Common-Emitter Stage

$$Z_i = \frac{h_{ib} + [h_{ib}h_{ob} + h_{rb}h_{fe}/(1+h_{fe})]Z_l}{h_{ob}Z_l + 1/(1+h_{fe})}$$

$$Z_o = \frac{h_{ib} + Z_g/(1+h_{fe})}{h_{ob}(h_{ib}+Z_g) + h_{rb}h_{fe}/(1+h_{fe})}$$

$$A_i = \frac{h_{fe}}{1 + h_{ob}Z_l(1+h_{fe})}$$

$$A_v = \frac{-h_{fe}Z_l}{h_{ib}(1+h_{fe})(1+h_{ob}Z_l) + h_{rb}h_{fe}Z_l}$$

$$G_t = \frac{4(h_{fe})^2 R_g}{R_l[h_{ib}(1+h_{fe})/R_l + R_g h_{ob}(1+h_{fe}) + R_g/R_l + h_{ib}h_{ob}(1+h_{fe}) + h_{rb}h_{fe}]^2}$$

$$G_m = \frac{(h_{fe})^2[h_{ib}h_{ob}(1+h_{fe}) + h_{rb}h_{fe}]}{[h_{ib}h_{ob}(1+h_{fe}) + h_{rb}h_{fe} + (1+h_{fe})\sqrt{(h_{ib}h_{ob})^2(1+h_{fe}) + h_{rb}h_{fe}}]^2}$$

$$R_{lm} = \sqrt{\frac{h_{ib}}{h_{ob}[h_{ib}h_{ob}(1+h_{fe}) + h_{rb}h_{fe}]}}$$

$$R_{im} = \sqrt{\frac{h_{ib}[h_{ib}h_{ob}(1+h_{fe}) + h_{rb}h_{fe}]}{h_{ob}}}$$

Table 8.3 Properties of the Terminated Common-Collector Stage

---

$$Z_i = \frac{h_{ib} + Z_l}{h_{ob}Z_l + 1/(1 + h_{fe})}$$

$$Z_o = \frac{h_{ib}(1 + h_{fe}) + Z_g}{(1 + h_{ob}Z_g)(1 + h_{fe})}$$

$$A_i = \frac{-(1 + h_{fe})}{1 + h_{ob}Z_l(1 + h_{fe})}$$

$$A_v = \frac{Z_l}{h_{ib} + Z_l}$$

$$G_t = \frac{4R_g}{R_l[1 + h_{ib}/R_l + R_g(h_{ob} + 1/R_l(1 + h_{fe})]^2}$$

$$G_m = \frac{1 + h_{fe}}{[1 + \sqrt{h_{ib}h_{ob}(1 + h_{fe})}]^2}$$

$$R_{lm} = \sqrt{\frac{h_{ib}}{h_{ob}(1 + h_{fe})}}$$

$$R_{im} = \sqrt{\frac{h_{ib}(1 + h_{fe})}{h_{ob}}}$$

8.6 The Degenerated Common-Emitter Stage

It is frequently desirable to use an unby-passed resistor in series with the emitter, for example, to increase the input impedance. The effect of adding such a resistor can readily be found by adding it to the h_{ib} terms in the equations of Table 8.2. Thus the input resistance under these conditions becomes:

$$Z_i = \frac{(h_{ib} + R_e) + [h_{ob}(h_{ib} + R_e) + h_{rb}h_{fe}/(1 + h_{fe})]Z_l}{h_{ob}Z_l + 1/(1 + h_{fe})} \tag{8.3}$$

If the load impedance is small, this simplifies to

$$Z_i = (h_{ib} + R_e)(1 + h_{fe}) \tag{8.4}$$

From this equation it is obvious that the effect of inserting resistance in the emitter lead is to produce an effective increase in input impedance of $(1 + h_{fe})$ times this resistance. Since this factor can be quite large, a very considerable increase in input impedance is possible by this method. Inspection of the other equations of Table 8.2 indicates that this resistor will also increase the output impedance, reduce the voltage amplification, transducer gain, and matched gain, but have no effect on current amplification.

8.7 Variation of Impedances, Amplification, and Gain with Generator and Load Resistances

The effects of varying generator and load resistances can be determined from the equations of Tables 8.1 through 8.3. Figure 8.5 shows the variation of input resistance of a 2N43 common-base stage as a function of load resistance, and the output resistance as a function of generator resistance. The input resistance is quite low, the output

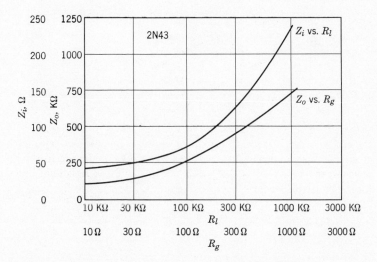

Fig. 8.5 Input and output impedances versus load and generator resistances, respectively, 2N43 in common-base configuration.

72 TRANSISTOR APPLICATIONS

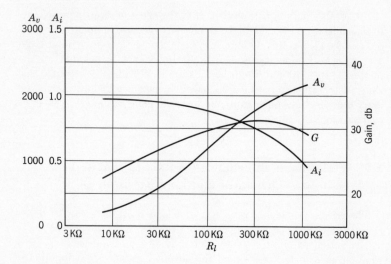

Fig. 8.6 Amplification and gain versus load resistance, 2N43 in common-base configuration.

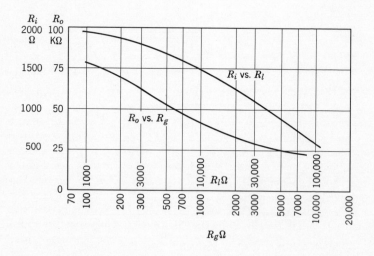

Fig. 8.7 Input and output resistances versus load and generator resistances, respectively, 2N43 in common-emitter configuration.

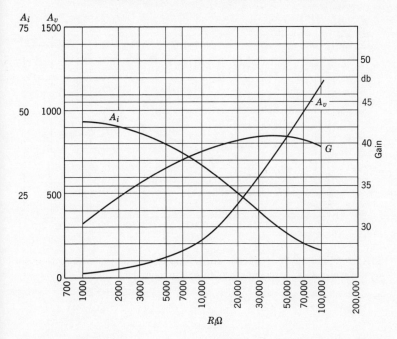

Fig. 8.8 Amplification and gain versus load resistance, 2N43 in common-emitter configuration.

resistance high. Figure 8.6 shows the variation of current and voltage amplification and of gain for this stage, as a function of load resistance.

Figures 8.7 and 8.8 show similar data on the variation of impedances, amplification, and gain for the common-emitter stage. In this configuration the input resistance is higher and the output resistance is lower, both by a factor of approximately h_{fe}, than for the common-base configuration. The gain is about 10 db higher for this configuration.

Figures 8.9 and 8.10 show the performance of the common-collector stage. It is interesting to note the linear relationship between impedances. This configuration has the highest input resistance and lowest output resistance. For this reason the so-called emitter-follower finds wide use where either high input impedance or low output impedance is desired. The direct dependence of these impedances on either the generator or load impedances must be remembered, however. The gain for

74 TRANSISTOR APPLICATIONS

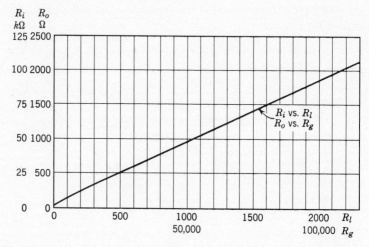

Fig. 8.9 Input and output resistances versus load and generator resistances, respectively, 2N43 in common-collector configuration.

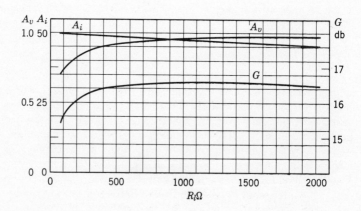

Fig. 8.10 Amplification and gain versus load resistance, 2N43 in common-collector configuration.

this configuration is the lowest of the three, and is quite flat over a wide range of load resistance.

REFERENCES

1. Shea, R. F. (Ed.), *Transistor Circuit Engineering*, Wiley, New York, 1957, pp. 73–79, 441–443.
2. Shea, R. F., *Transistor Audio Amplifiers*, Wiley, New York, 1955, pp. 72–95.

9

Cascaded Stages

9.1 Introduction

In the previous chapter the single transistor stage was discussed and equations were presented whereby the impedances, amplifications, and gains could be computed. In this chapter the combination of more than one stage in cascade are discussed, and the special problems presented as a result of this coupling will be described.

Cascading electron tube stages normally presents no special problems, except at high frequencies, since the tube is essentially a unilateral device. The transistor, on the other hand, is not unilateral, hence each stage produces a loading effect on both the preceding stage and on the following stage. This loading can, for example, considerably complicate the problem of aligning a high-frequency transistor amplifier, and even at relatively low frequencies the effect of the adjacent stages can be quite pronounced.

The finite loading presented by transistor amplifiers further complicates the design of the coupling network between stages. An additional factor which must be considered is the effect of emitter degeneration. As pointed out in the previous chapter, an unby-passed resistor in the emitter lead acts as though it were inserted in series with the base, multiplied by the factor $1 + h_{fe}$ approximately. Where the resistor is by-passed by a capacitor, the impedance introduced into the base circuit will be frequency-dependent. At low frequencies it will approach the resistance multiplied by $1 + h_{fe}$, whereas at high frequencies the effect will disappear. The effect is, therefore, somewhat similar to the insertion of an added coupling capacitor into the base lead.

The above effects will now be discussed in reasonable detail.

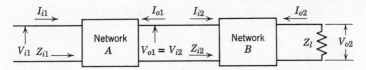

Fig. 9.1 Block diagram of two cascaded stages, showing voltages and currents.

9.2 Basic Relationships in Cascaded Networks

Figure 9.1 shows in block-diagram form the elements of two cascaded stages, A and B. In this figure they are indicated as being networks, in order that the treatment may be general. Thus, the two networks may both be active, e.g., two transistor stages, or one or both may be passive, e.g., one may be the coupling network following or preceding an active transistor stage.

As shown on the figure, network B is terminated in a load impedance Z_l, across which an output voltage V_{o2} appears. The output current I_{o2} flows through this load impedance, and, using our standard convention whereby currents flow *into* the network, $V_{o2} = -Z_l I_{o2}$. Thus loaded, network B presents an input impedance Z_{i2}, the value of which can be obtained from the tables given in the previous chapter. The input voltage to network B is V_{i2} and the input current is I_{i2}.

Network A is now loaded with an output impedance equal to the input impedance presented by network B, thus $Z_{l1} = Z_{i2}$. Furthermore, $V_{o1} = V_{i2}$ and $I_{o1} = -I_{i2}$ (because of the opposite directions). Network A will now have an input impedance Z_{i1}, an input current I_{i1}, and an input voltage V_{i1}.

Each network may be analyzed individually and the results combined, or the tandem may be treated as one composite network by using the equations given in Section 2.5 of Chapter 2. In this chapter most of the emphasis will be on the individual networks and their effects on their neighbors.

9.3 Input Impedance of the Cascaded Pair

The input impedance of stage B can be obtained from the equations for the terminated network, and is:

$$Z_{i2} = \frac{(h_{11})_2 + (\Delta^h)_2 Z_l}{1 + (h_{22})_2 Z_l} \tag{9.1}$$

In the above and the following equations the subscript outside the parameter parentheses indicates the respective networks, 1 for network A, 2 for B.

This impedance, as noted above, becomes the load impedance, in turn, for the first network. Substituting this for Z_{l1} and solving, we obtain, for the input impedance of the first network, and thus of the complete cascade:

$$Z_{i1} = \frac{(h_{11})_1[1 + (h_{22})_2 Z_l] + (\Delta^h)_1[(h_{11})_2 + (\Delta^h)_2 Z_l]}{[1 + (\Delta^h)_2][1 + (h_{22})_2 Z_l]} \qquad (9.2)$$

$$= \frac{(h_{11})_1}{1 + (\Delta^h)_2} + \frac{(\Delta^h)_1[(h_{11})_2 + (\Delta^h)_2 Z_l]}{[1 + (\Delta^h)_2][1 + (h_{22})_2 Z_l]} \qquad (9.3)$$

From this equation it is evident that the input impedance of a cascaded pair will be approximately equal to the h_{11} of the first network if the second network consists of either a common-base or common-emitter stage, since the determinants are low for either of these arrangements. Typically, for a common-emitter second stage, the second term of eq. 9.3 will approach about 10 ohms as Z_l approaches infinity, thus the effect is negligible if the first stage is also a common-emitter. On the other hand, if the second stage is a common-collector stage, the second term becomes predominant since the determinants are quite large, thus the cascaded emitter-follower can present a very high input impedance, up to the megohm range.

9.4 Current Amplification of the Cascaded Pair

The current amplification A_{i2} of the second stage is

$$A_{i2} = \frac{(h_{21})_2}{1 + (h_{22})_2 Z_l} \qquad (9.4)$$

Substituting the value of load impedance given by eq. 9.1, the current amplification of the first stage is

$$A_{i1} = \frac{(h_{21})_1[1 + (h_{22})_2 Z_l]}{1 + (h_{22})_2 Z_l + (h_{22})_1[(h_{11})_2 + (\Delta^h)_2 Z_l]} \qquad (9.5)$$

The product of the above two current amplifications is that of the cascaded pair.

$$A_{i12} = \frac{(h_{21})_1 (h_{21})_2}{1 + (h_{22})_2 Z_l + (h_{22})_1[(h_{11})_2 + (\Delta^h)_2 Z_l]} \qquad (9.6)$$

9.5 Voltage Amplification of the Cascaded Pair

The voltage amplification A_{v2} of the second stage is

$$A_{v2} = \frac{-(h_{21})_2 Z_l}{(h_{11})_2 + (\Delta^h)_2 Z_l} \tag{9.7}$$

Substituting the value of load impedance of the first stage, its voltage amplification is

$$A_{v1} = \frac{-(h_{21})_1[(h_{11})_2 + (\Delta^h)_2 Z_l]}{(h_{11})_1[1 + (h_{22})_2 Z_l] + (\Delta^h)_1[(h_{11})_2 + (\Delta^h)_2 Z_l]} \tag{9.8}$$

The voltage amplification for the cascaded pair is the product of the individual amplifications

$$A_{v12} = \frac{(h_{21})_1 (h_{21})_2 Z_l}{(h_{11})_1[1 + (h_{22})_2 Z_l] + (\Delta^h)_1[(h_{11})_2 + (\Delta^h)_2 Z_l]} \tag{9.9}$$

9.6 Power Gain of the Cascaded Pair

The power gain is obtained as the product of the over-all current and voltage amplifications (taken as a magnitude), and hence can be obtained from eqs. 9.6 and 9.9. The equations for transducer and matched gain are extremely unwieldy, however; if desired, they may be obtained by using the equations of Section 2.5 of Chapter 2 to get one composite set of parameters, then inserting them in the appropriate equations of Chapter 8. Alternatively, one can insert specific values of load impedances and determine these gains for each stage.

9.7 Combinations of Transistor Configurations

It is occasionally desirable to combine mixed configurations, e.g., a common-collector stage (emitter-follower) driving a common-emitter stage, or vice versa. The former provides high input impedance at some sacrifice of gain, the latter low driving impedance, and under certain conditions such terminal impedances may be necessary. Table 9.1 gives the over-all transducer gain of mixed pairs of stages, using the 2N43 transistors.

*Table 9.1 Transducer Gain of Mixed Combinations—
Two-Stage Amplifier (2N43)*

1st Stage:		CE	CB	CC	CE	CB	CE	CC	CC	CB
2nd Stage:		CE	CB	CC	CB	CE	CC	CE	CB	CC

R_g	R_l									
100	100	46.9	2.5	3.0	13.2	36.6	46.6	13.6	7.5	36.6
1,000	1,000	63.3	5.1	5.6	30.1	38.8	57.1	33.4	25.8	38.5
10,000	10,000	68.0	5.4	5.9	38.0	36.1	44.8	51.5	36.7	33.7
100,000	100,000	57.6	4.6		38.6	24.1		55.3	38.5	18.9
100	1,000	56.5	12.5	−4.4	23.2	46.2	52.6	23.5	17.5	44.9
100	10,000	63.5	22.5	−14	33.1	53.2	49.3	32.2	27.1	46.2
100	100,000	61.5	31.7		42.4	52.5		37.0	33.4	38.9
100,000	100	42.6	−24.6	30.2	9.4	9.6	41.0	37.2	9.5	9.7
10,000	100	51.3	−14.6	22.7	18.0	19.6	49.8	32.7	16.9	19.7
1,000	100	53.6	−4.9	13.0	20.1	29.2	52.9	23.5	15.8	29.4

In Table 9.1, the effect of the coupling and biasing networks have been omitted; in other words, it has been assumed that the only load the first stage sees is that of the input impedance of the second stage. These gains will not, therefore, be realizable in practice; however, they indicate the relative advantages and disadvantages of various combinations.

Note that in all cases the gain obtained with the cascaded common-emitter stages is highest, regardless of the value of generator or load resistance. However, some other combinations approach this figure, e.g., the *CE* input stage followed by the *CC* second stage is only 0.3 db below the *CE-CE* combination when working from a source impedance of 100 ohms into a load of 100 ohms. Similarly, the *CC-CE* combination is only 2.3 db below the *CE-CE* combination for 100,000-ohm source and load. Furthermore, there may be other features making these other combinations desirable, despite the loss in gain. For example, some phonograph pickups require high input impedance for proper frequency response. This may be obtained by using the *CC* stage as the first stage. As another example, a low output impedance may be required to provide proper damping for the load. This may dictate the *CC* output stage. To summarize, there is no fixed rule that dictates that any particular combination is best—each must be considered individually.

CASCADED STAGES 81

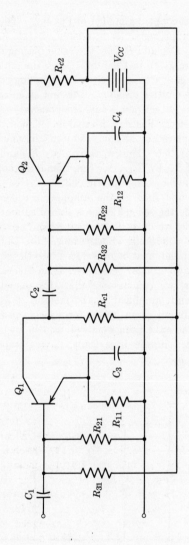

Fig. 9.2 Circuit of two-stage RC-coupled amplifier.

9.8 The Coupling Network

Consider the two-stage RC-coupled amplifier of Fig. 9.2. This shows an output transistor Q_2 supplying a load resistor R_{c2} through the supply V_{CC}. Resistors R_{12}, R_{22}, and R_{32} are the usual biasing resistors for which the equations were given in Chapter 7, with the added subscripts indicating second stage. Capacitor C_2 is the coupling capacitor and C_4 serves to by-pass the emitter resistor. The first stage is similar in arrangement, and uses the subscript 1 for its components.

It is assumed that the battery has zero internal impedance, thus the two bias resistors may be replaced by an equivalent one, having a value equal to that of the two in parallel. Thus we can replace R_{21} and R_{31} by one resistor R_{b1} and R_{22} and R_{32} by R_{b2}. We can, therefore, replace the circuit of Fig. 9.2 by the simpler version of Fig. 9.3.

The properties of the second stage of this amplifier are defined by eqs. 9.1, 9.4, and 9.7. Since the second stage is shown as a common-emitter stage in this example, we must use the corresponding common-emitter parameters in the equations. Furthermore, the impedance formed by R_{12} and C_4 in parallel must be introduced into the equations by adding this impedance to the proper parameter, h_{ib} in this case. As an approximation, however, we can assume that this impedance is reflected into the input circuit, amplified by the factor $1 + h_{fe}$. We can, therefore, further modify our circuit to the form shown in Fig. 9.4. Here the emitter impedances have been removed, and in their places networks have been inserted in the base leads, having impedances increased by the factor $1 + h_{fe}$.

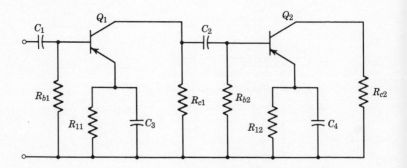

Fig. 9.3 Simplified representation of RC-coupled amplifier.

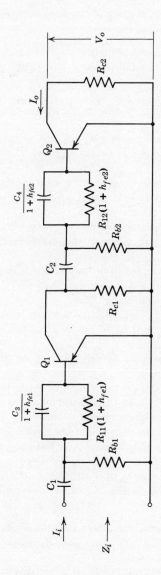

Fig. 9.4 Illustrating the transfer of emitter impedance to base circuit.

84   TRANSISTOR APPLICATIONS

The equations for the over-all amplifier containing all these elements become very unwieldy and no attempt will be made to develop them here. Instead, the procedure for obtaining the over-all performance will be indicated, and an idea will be given of the relative magnitude of some of the effects of the coupling elements.

1. The input impedance and current amplification of the second stage are obtained from eqs. 9.1 and 9.4, using the common-emitter parameters. Alternatively, the equations given in Table 8.2 of the last chapter may be used.
2. This input impedance is added to that of the shunt mesh in the second transistor base.
3. This impedance is now shunted by the effective bias resistance, R_{b2}. The result is the total impedance to the right of the coupling capacitor C_2.
4. The reactance of C_2 is now added, and then the whole is shunted by R_{c1}. The result is the effective load impedance seen by transistor Q_1.
5. Using this impedance as the load for the first stage, its input impedance and current amplification are now computed, in the same manner as for the second stage in procedure 1.
6. Again, the base lead mesh is added to the input impedance, and the whole is shunted by R_{b1} and added to the reactance of C_1 to give the input impedance of the over-all amplifier.
7. The input current I_i divides between R_{b1} and the total impedance to the right of this point inversely as impedance. Thus, the portion of I_i flowing into Q_1 is determined.
8. This is multiplied by the current amplification of the first stage, A_{i1}, as obtained in procedure 5.
9. This output current of Q_1 divides between R_{c1} and the impedance to its right and again between R_{b2} and the impedance to the right of that resistor. This gives the input current to transistor Q_2.
10. This current is multiplied by the current amplification A_{i2}, obtained in procedure 1 to give the output current I_o. The output voltage will be $-R_{c2}I_o$.
11. The input voltage is $Z_i I_i$, hence the over-all voltage amplification is $-R_{c2}I_o/Z_i I_i = -A_{i12}R_{c2}/Z_i$, where A_{i12} is the over-all current amplification.

The over-all power gain will be the power into the load, $R_{c2}(I_o)^2$ divided by the input power $R_i(I_i)^2$ or $[(A_{i12})^2 R_{c2}]/R_i$, where R_i is the real part of the total input impedance Z_i.

The transducer gain is the output power $R_{c2}(I_o)^2$ divided by the available power from the source, $P_{av} = V_g^2/4R_g = [I_i^2(Z_g + Z_i)^2]/4R_g$. Thus the transducer gain of the complete two-stage amplifier is

$$G_t = \frac{4R_g R_{c2}}{(Z_g + Z_i)^2}(A_{i12})^2 \qquad (9.10)$$

In all the above steps it must be remembered that the impedances become complex at those low frequencies where the capacitor reactances are appreciable, hence all the additions and multiplications implied above are done by using complex numbers.

9.9 Effects of the Coupling Capacitor and Emitter By-Pass

As indicated above, these capacitors produce considerable effect at low frequencies. The effects are not easy to evaluate because the increasing reactance of the coupling capacitor, for example, attenuates the portion of collector current supplied to the following base, but it also increases the collector current by increasing the value of collector impedance, thus the two effects partially compensate.

Figure 9.5 shows the effect of the coupling capacitor in reducing the low-frequency gain of a typical RC-coupled amplifier. Obviously values in excess of 10 μfd are necessary if true low-frequency reproduction is to be achieved.

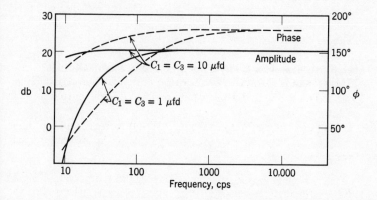

Fig. 9.5 Effect of coupling capacitor on frequency response.

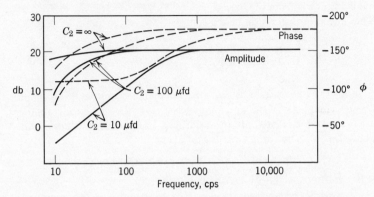

Fig. 9.6 Effect of emitter by-pass capacitor on frequency response.

Figure 9.6 shows the effect of the capacitor across the emitter resistor, in this case 1000 ohms. The very serious effect of inadequate by-passing is very evident and indicates that values in excess of 100 μfd are required to obtain adequate low-frequency response. Fortunately, the voltages encountered in transistor amplifiers are very low, consequently tantalytic capacitors can be used which combine high values of capacitance with extremely small size and low leakage. Another expedient frequently encountered is the use of breakdown diodes to provide the emitter bias. These diodes have low dynamic impedance and thus obviate the necessity of using large capacitors.

REFERENCE

1. Shea, R. F., *Transistor Audio Amplifiers*, Wiley, New York, 1955.

10

Feedback Networks— Matrix Transposition

10.1 Introduction

In the opening chapters the fundamentals of network theory and the use of the various forms of matrix representations was presented. In this chapter, a number of examples will be given, illustrating the application of these powerful tools to the solution of typical feedback circuits. In particular, the use of the transposition matrix frequently permits analysis of complicated circuits involving feedback loops, which would be extremely difficult to solve otherwise.

10.2 Solution of the Degenerated Emitter Circuit by Matrices

We have previously analyzed the effect of inserting an impedance in the emitter lead by means of its effects directly on the parameters of the transistor (see Section 8.6, Chapter 8). This arrangement can also be analyzed by considering the transistor and the emitter impedance to be two separate networks, connected in series (see Section 2.1, Chapter 2). Figure 10.1 illustrates this approach. The transistor is considered to be one network, as indicated by the upper dotted lines, the emitter impedance Z_e the other network, shown as the lower dotted enclosure. As indicated by the connecting leads, these two networks are connected in series on input and output, hence, as shown in Chapter 2, we can analyze this combination by taking the z parameters of each network and adding them together on a term-by-term basis. Performing this operation, we

88 TRANSISTOR APPLICATIONS

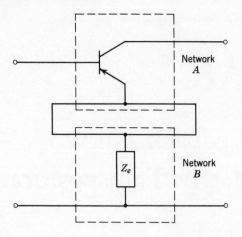

Fig. 10.1 Transistor with emitter impedance, shown as two matrices in series.

obtain, as the Z matrix of the combination,

$$[Z] = \begin{bmatrix} \dfrac{\Delta^h_e}{h_{oe}} + Z_e; & \dfrac{h_{re}}{h_{oe}} + Z_e \\ -\dfrac{h_{fe}}{h_{oe}} + Z_e; & \dfrac{1}{h_{oe}} + Z_e \end{bmatrix} \quad (10.1)$$

Converting to the common-base specification sheet parameters (with the exception of h_{fe}),

$$[Z] = \begin{bmatrix} h_{ib} + \dfrac{h_{rb}}{h_{ob}}\left(\dfrac{h_{fe}}{1 + h_{fe}}\right) + Z_e; & h_{ib} - \dfrac{h_{rb}}{h_{ob}(1 + h_{fe})} + Z_e \\ \dfrac{-h_{fe}}{h_{ob}(1 + h_{fe})} + Z_e; & \dfrac{1}{h_{ob}(1 + h_{fe})} + Z_e \end{bmatrix} \quad (10.2)$$

If it is now desired to convert from the above z parameters to h parameters, we can use the equivalences:

$$[H] = \begin{bmatrix} \dfrac{\Delta^z}{z_{22}} & \dfrac{z_{12}}{z_{22}} \\ -\dfrac{z_{21}}{z_{22}} & \dfrac{1}{z_{22}} \end{bmatrix} \quad (10.3)$$

From eqs. 10.2, we compute $\Delta^z = z_{11}z_{22} - z_{12}z_{21}$:

$$\Delta^z = \frac{1}{h_{ob}}[h_{ib} + (1 + h_{rb})Z_e] \tag{10.4}$$

and, solving the equations in the above matrix, we obtain:

$$h_{ie}' = \frac{h_{ib} + (1 + h_{rb})Z_e}{h_{ob}Z_e + 1/(1 + h_{fe})} \tag{10.5}$$

$$h_{re}' = \frac{h_{ib}h_{ob}(1 + h_{fe}) - h_{rb} + h_{ob}(1 + h_{fe})Z_e}{1 + h_{ob}(1 + h_{fe})Z_e} \tag{10.6}$$

$$h_{fe}' = \frac{h_{fe} - h_{ob}(1 + h_{fe})Z_e}{1 + h_{ob}(1 + h_{fe})Z_e} \tag{10.7}$$

$$h_{oe}' = \frac{h_{ob}}{h_{ob}Z_e + 1/(1 + h_{fe})} \tag{10.8}$$

10.3 Collector-to-Base Feedback

This arrangement is frequently used to provide phase compensation and prevent oscillation at high frequencies. Figure 10.2 illustrates the connection of an impedance Z_f between collector and base. Z_f may incorporate a blocking capacitor, thus providing only a-c feedback, or the capacitor may be omitted, giving d-c feedback as well. The

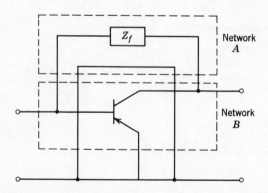

Fig. 10.2 Transistor with shunt feedback resistor, shown as two matrices in parallel.

capacitor may have low impedance, being provided only for d-c isolation, or it may have appreciable reactance at the frequency of interest, producing appreciable phase shift.

By adding a fictitious ground lead as shown in Fig. 10.2, we can show the effect of two networks connected with their inputs and outputs in parallel. Again referring to the earlier chapters, we see that this condition requires use of the Y matrices, added term by term. The Y parameters for the transistor are:

$$[Y_t] = \begin{bmatrix} \dfrac{1}{h_{ib}(1+h_{fe})}; & \dfrac{h_{rb}}{h_{ib}(1+h_{fe})} - h_{ob} \\ \dfrac{h_{fe}}{h_{ib}(1+h_{fe})}; & h_{ob} + \dfrac{h_{rb}h_{fe}}{h_{ib}(1+h_{fe})} \end{bmatrix} \quad (10.9)$$

For the feedback network the matrix elements are:

$$y_{11} = y_{22} = Y_f = \frac{1}{Z_f}$$

$$y_{12} = y_{21} = -Y_f = -\frac{1}{Z_f} \quad (10.10)$$

Combining the two:

$$[Y''] = \begin{bmatrix} \dfrac{1}{h_{ib}(1+h_{fe})} + Y_f; & \dfrac{h_{rb}}{h_{ib}(1+h_{fe})} - h_{ob} - Y_f \\ \dfrac{h_{fe}}{h_{ib}(1+h_{fe})} - Y_f; & \dfrac{h_{rb}h_{fe}}{h_{ib}(1+h_{fe})} + h_{ob} + Y_f \end{bmatrix} \quad (10.11)$$

Again, we can convert to the h parameters:

$$[H''] = \begin{bmatrix} \dfrac{1}{y_{11}} & \dfrac{-y_{12}}{y_{11}} \\ \dfrac{y_{21}}{y_{11}} & \dfrac{\Delta^y}{y_{11}} \end{bmatrix} \quad (10.12)$$

From eq. 10.11,

$$\Delta^y = \frac{h_{ob} + (1+h_{rb})Y_f}{h_{ib}} \quad (10.13)$$

and from eqs. 10.11 and 10.13, using the relationships of eq. 10.12,

$$h_{ie}'' = \frac{h_{ib}(1+h_{fe})}{1+h_{ib}(1+h_{fe})Y_f} \quad (10.14)$$

$$h_{re}'' = \frac{h_{ib}h_{ob}(1 + h_{fe}) - h_{rb} + h_{ib}(1 + h_{fe})Y_f}{1 + h_{ib}(1 + h_{fe})Y_f} \quad (10.15)$$

$$h_{fe}'' = \frac{h_{fe} - h_{ib}(1 + h_{fe})Y_f}{1 + h_{ib}(1 + h_{fe})Y_f} \quad (10.16)$$

$$h_{oe}'' = \frac{(1 + h_{fe})[h_{ob} + (1 + h_{rb})Y_f]}{1 + h_{ib}(1 + h_{fe})Y_f} \quad (10.17)$$

10.4 Use of the Transfer and Transposition Matrices

Frequently analysis of a circuit is complicated by feedback meshes which do not have a common ground connection with the main amplifier circuit. In such cases, the use of the following technique may prove of considerable value. It will be demonstrated first on a relatively simple circuit, then its applicability to a more complicated circuit will be shown.

Figure 10.3 shows an elementary transistor stage, connected in the common-base configuration in Fig. 10.3a and in the common-emitter configuration in 10.3b. The input and output voltages are designated by the symbols V in the former, by E in the latter, and the currents are indicated by the symbols I and J respectively. The four terminals carry the same numbers.

We can express the relationships between currents and voltages for these two circuits as follows, using the matrix notation:

$$[V] = [Z_b][I] \quad \text{and} \quad [E] = [Z_e][J] \quad (10.18)$$

where $[Z_b]$ represents the matrix for the common-base configuration

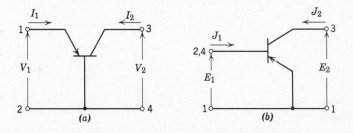

Fig. 10.3 Transposition of common-base stage to common-emitter stage.

$(z_{ib}, z_{rb}, z_{fb},$ and $z_{ob})$, and $[Z_e]$ represents the common-emitter z parameters.

From inspection we see that $E_1 = -V_1$ and $E_2 = V_2 - V_1$. These relationships can be put in the form of two simultaneous equations with coefficients of 0, 1, or -1:

$$E_1 = -1V_1 + 0V_2$$
$$E_2 = -1V_1 + 1V_2$$

or, in matrix form:

$$[E] = \begin{bmatrix} -1 & 0 \\ -1 & 1 \end{bmatrix} [V] \qquad (10.19)$$

The matrix $\begin{bmatrix} -1 & 0 \\ -1 & 1 \end{bmatrix}$ is the transfer matrix from the common-base configuration to the common-emitter. It can be shown that the relationship between the two impedance matrices is as follows:

$$[Z_e] = \begin{bmatrix} -1 & 0 \\ -1 & 1 \end{bmatrix} [Z_b] \begin{bmatrix} -1 & -1 \\ 0 & 1 \end{bmatrix} \qquad (10.20)$$

The first numerical matrix above is the transfer matrix, as given in eq. 10.19. The second is its transposition, obtained by interchanging the rows and columns of the transfer matrix (simply interchange the 12 and the 21 elements of the matrix in a second-order one).

Substituting the z parameters now,

$$[Z_e] = \begin{bmatrix} -1 & 0 \\ -1 & 1 \end{bmatrix} \begin{bmatrix} z_{ib} & z_{rb} \\ z_{fb} & z_{ob} \end{bmatrix} \begin{bmatrix} -1 & -1 \\ 0 & 1 \end{bmatrix} \qquad (10.21)$$

The mechanism for performing this multiplication is to premultiply the Z_b matrix first by the transfer matrix, then postmultiply the result by the transposition matrix:

$$\begin{bmatrix} -1 & 0 \\ -1 & 1 \end{bmatrix} \begin{bmatrix} z_{ib} & z_{rb} \\ z_{fb} & z_{ob} \end{bmatrix} = \begin{bmatrix} -z_{ib}; & -z_{rb} \\ -z_{ib} + z_{fb}; & -z_{rb} + z_{ob} \end{bmatrix} \qquad (10.22)$$

$$\begin{bmatrix} -z_{ib}; & -z_{rb} \\ -z_{ib} + z_{fb}; & -z_{rb} + z_{ob} \end{bmatrix} \begin{bmatrix} -1 & -1 \\ 0 & 1 \end{bmatrix}$$

$$= \begin{bmatrix} z_{ib}; & z_{ib} - z_{rb} \\ z_{ib} - z_{fb}; & z_{ib} - z_{fb} - z_{rb} + z_{ob} \end{bmatrix} \qquad (10.23)$$

We have thus obtained the common-emitter z parameters, in terms of the common-base:

$$z_{ie} = z_{ib} \qquad z_{re} = z_{ib} - z_{rb} \qquad z_{fb} = z_{ib} - z_{fb}$$
$$z_{oe} = z_{ib} - z_{fb} - z_{rb} + z_{ob}$$

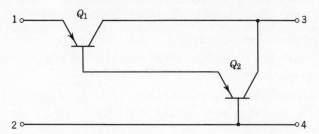

Fig. 10.4 The common-base Darlington configuration.

10.5 Analysis of Darlington Pair Using Transposition Matrix

A more exotic application of this technique is illustrated by the circuit of Fig. 10.4, the familiar Darlington pair, shown in its common-base configuration. It is desired to obtain the parameters of an equivalent single transistor.

The first step is to reorient the circuit until it assumes some more workable aspect. Rotating it as shown in Fig. 10.5 causes it to appear as a cascade of a pair of common-collector stages. Following the method used in the first example, we will perform the following steps:

1. Determine the Z matrix of the circuit of Fig. 10.5.
2. Determine the transfer matrix representing the relationship between the input and output voltages of the two circuits.

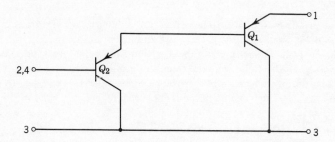

Fig. 10.5 The common-emitter Darlington configuration.

94   TRANSISTOR APPLICATIONS

3. Premultiply the Z matrix by this transfer matrix, then post-multiply by its transposition. This will give us the Z matrix of the Darlington pair.
4. If desired, the Z matrix can now be transformed to any of the other forms, e.g., the H matrix, by the table of interrelations.

As indicated above, the first step is to obtain the Z matrix of the circuit of Fig. 10.5. Assuming we have the h parameters of the individual transistors to start with, we may proceed in either of two ways: (1) we may convert our h parameters to z parameters, using the tables of parameter interrelationships, then use eqs. 2.21 through 2.24 of Chapter 2 to derive the z parameters of the cascaded pair; (2) we can convert the h parameters of the individual transistors to a parameters, using again the table of interrelations, then multiply the two A matrices to obtain an A matrix of the pair, then finally reconvert back to h parameters. We will choose the latter method in this example.

The A matrix of a common-collector stage is obtained from the H matrix of the stage, in terms of the spec parameters:

$$[H_c] = \frac{1}{1+h_{fb}} \begin{bmatrix} h_{ib}; & 1+h_{fb} \\ -1;\cdot & h_{ob} \end{bmatrix} \qquad \Delta^h{}_c = \frac{1}{1+h_{fb}} \qquad (10.24)$$

Converting $1 + h_{fb}$ to $1/(1+h_{fe})$:

$$[H_c] = (1+h_{fe}) \begin{bmatrix} h_{ib} & 1/(1+h_{fe}) \\ -1 & h_{ob} \end{bmatrix} \qquad \Delta^h{}_c = 1+h_{fe} \qquad (10.25)$$

The A matrix is:

$$[A_c] = \begin{bmatrix} -\dfrac{\Delta^h}{h_{21}}; & -\dfrac{h_{11}}{h_{21}} \\ -\dfrac{h_{22}}{h_{21}}; & \dfrac{-1}{h_{21}} \end{bmatrix} = \begin{bmatrix} 1; & h_{ib} \\ h_{ob}; & \dfrac{1}{1+h_{fe}} \end{bmatrix} \qquad (10.26)$$

The A matrix is thus seen to have a very simple form, when expressed in terms of the common-base parameters and h_{fe}.

The over-all A matrix is obtained by multiplying the two individual matrices. In the following equations the number following the letters in the subscript indicates whether the parameter pertains to Q_1 or to Q_2 (note that the sequence of transistors has been reversed in the rotation).

$$[A_{cc}] = \begin{bmatrix} 1 + h_{ib2}h_{ob1} & h_{ib1} + \dfrac{h_{ib2}}{1+h_{fe1}} \\ h_{ob2} + \dfrac{h_{ob1}}{1+h_{fe2}} & h_{ib1}h_{ob2} + \dfrac{1}{(1+h_{fe1})(1+h_{fe2})} \end{bmatrix} \qquad (10.27)$$

For simplicity, let us designate the terms of the above matrix as follows:

$$A_{11} = 1 + h_{ib2}h_{ob1}$$

$$A_{12} = h_{ib1} + \frac{h_{ib2}}{1 + h_{fe1}}$$

$$A_{21} = h_{ob2} + \frac{h_{ob1}}{1 + h_{fe2}} \tag{10.28}$$

$$A_{22} = h_{ib1}h_{ob2} + \frac{1}{(1 + h_{fe1})(1 + h_{fe2})}$$

The Z matrix of the cascaded common-collector stage for which we have now obtained the A matrix is, in terms of the above parameters,

$$[Z_{cc}] = \begin{bmatrix} \dfrac{A_{11}}{A_{21}} & \dfrac{\Delta^A}{A_{21}} \\ \dfrac{1}{A_{21}} & \dfrac{A_{22}}{A_{21}} \end{bmatrix} \tag{10.29}$$

We now transpose this impedance matrix to that of the desired Darlington circuit. The input and output voltages are:

Cascaded common-collector: $V_{\text{in}} = V_{2-3}$ $\quad V_{\text{out}} = V_{1-3}$
Darlington common-base: $V_{\text{in}} = V_{1-2}$ $\quad V_{\text{out}} = V_{3-4} = V_{3-2}$

Thus, the matrix relationship between circuits, or the transfer matrix is:

$$[V_D] = \begin{bmatrix} -1 & 1 \\ -1 & 0 \end{bmatrix} [V_{cc}] \tag{10.30}$$

The Z matrix of the C. B. Darlington is, therefore,

$$[Z_D] = \begin{bmatrix} -1 & 1 \\ -1 & 0 \end{bmatrix} [Z_{cc}] \begin{bmatrix} -1 & -1 \\ 1 & 0 \end{bmatrix} \tag{10.31}$$

Performing the above series of multiplications on the parameters given in eq. 10.29, we obtain:

$$Z_D = \frac{1}{A_{21}} \begin{bmatrix} A_{11} - 1 - \Delta^A + A_{22}; & A_{11} - 1 \\ A_{11} - \Delta^A; & A_{11} \end{bmatrix} \tag{10.32}$$

The determinant Δ^A is calculated to be:

$$\Delta^A = \frac{1}{(1 + h_{fe1})(1 + h_{fe2})} - \frac{h_{ib1}h_{ob1}}{1 + h_{fe2}} - \frac{h_{ib2}h_{ob2}}{1 + h_{fe1}}$$
$$+ h_{ib1}h_{ib2}h_{ob1}h_{ob2} \tag{10.33}$$

96 TRANSISTOR APPLICATIONS

If desired, the z parameters of the Darlington pair may now be obtained by substituting the values of the A parameters and of the determinant from eqs. 10.28 and 10.33 in the matrix terms of eq. 10.32.

It is informative to obtain the h parameters of the Darlington pair. This may be accomplished by using the Z to H interrelation:

$$[H_D] = \begin{bmatrix} \dfrac{\Delta^z}{z_{22}}; & \dfrac{z_{12}}{z_{22}} \\ -\dfrac{z_{21}}{z_{22}}; & \dfrac{1}{z_{22}} \end{bmatrix} \qquad (10.34)$$

Substituting the appropriate values of the Z_D parameters in this matrix, then using the eqs. 10.28 to convert these to the common-base parameters and h_{fe}, we obtain, for the Darlington pair,

$$(h_{ib})_D = \frac{h_{ib1} + h_{ib2}/(1 + h_{fe1})}{1 + h_{ib2}h_{ob1}} \qquad (10.35)$$

$$(h_{rb})_D = \frac{h_{ib2}h_{ob1}}{1 + h_{ib2}h_{ob1}} \qquad (10.36)$$

$$(h_{fb})_D = \frac{-1 - h_{ib2}h_{ob1} + 1/(1 + h_{fe1})(1 + h_{fe2})}{1 + h_{ib2}h_{ob1}} + \frac{-h_{ib1}h_{ob1}/(1 + h_{fe2}) - h_{ib2}h_{ob2}/(1 + h_{fe1}) + h_{ib1}h_{ib2}h_{ob1}h_{ob2}}{1 + h_{ib2}h_{ob1}} \qquad (10.37)$$

$$(h_{ob})_D = \frac{h_{ob2} + h_{ob1}/(1 + h_{fe2})}{1 + h_{ib2}h_{ob1}} \qquad (10.38)$$

The above equations may be simplified by neglecting the relatively minor terms:

$$(h_{ib})_D \simeq h_{ib1} \qquad (10.39)$$

$$(h_{rb})_D \simeq h_{ib2}h_{ob1} \qquad (10.40)$$

$$(h_{fb})_D \simeq h_{fb1} + h_{fb2} + h_{fb1}h_{fb2} \qquad (10.41)$$

$$(h_{ob})_D \simeq h_{ob2} \qquad (10.42)$$

We can also calculate the equivalent h_{fe} of the Darlington pair:

$$(h_{fe})_D = h_{fe1} + h_{fe2} + h_{fe1}h_{fe2} \qquad (10.43)$$

11
D-C Amplifiers

11.1 Introduction

In this chapter a number of techniques will be described which are commonly used for d-c amplification, including direct-coupled amplifiers, differential amplifiers, chopper-type amplifiers, modulated-carrier amplifiers (of which the chopper-type is also one), and field-effect transistor amplifiers. It is impractical to present more than the basic elements of these techniques herein, as there are countless variations of each. Instead, typical designs will be illustrated and the basic features of each technique will be discussed.

11.2 Direct-Coupled Amplifiers

Figure 11.1 shows a simple direct-coupled two-stage amplifier. The resistance values have been chosen to produce the voltages indicated with collector currents of 250 μa and 1 ma respectively. There are a number of obvious deficiencies in this circuit:

1. The input voltage is not zero under quiescent conditions, but is 0.9 v above zero, also there is a finite input current under this condition, 5 μa. Thus the input cannot be connected directly to a low-impedance source without upsetting the operating biases, and must be connected to a bucking voltage if the true input level is to be zero.
2. Similarly, the output is nonzero, and some means must be employed to buck this down if a normally zero output is desired.

98   TRANSISTOR APPLICATIONS

3. Relatively large resistors are introduced into both emitters, producing considerable loss in gain because of the degenerative effects of these resistors.
4. The collector-base voltage of the first transistor is limited, and in this case is only 1.7 v.
5. There is no provision for over-all feedback, hence the operating biases will be subject to considerable shift with transistor interchange and with temperature.

These adverse features of this type of direct-coupled amplifier can be alleviated, and in many cases completely eliminated, by a number of expedients, such as using a negative supply voltage, as well as the positive supply shown in Fig. 11.1; using breakdown diodes in place of resistors, which produce the same voltage drops, but have low dynamic impedance and, therefore, do not produce as much degeneration; combining p-n-p and n-p-n transistors in complementary arrangements, and coupling the output stage back to the input to produce d-c feedback. Figure 11.2 illustrates the application of most of the above expedients. A combination of n-p-n and p-n-p silicon transistors is used in complementary fashion; two breakdown diodes are used—the 1N429 to establish emitter bias for the 2N328A without introducing degeneration, the other to set the emitter bias of the first transistor properly for zero input voltage and to avoid a-c feedback (this amplifier can also be used for an

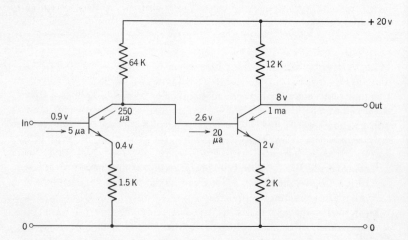

Fig. 11.1   Direct-coupled two-stage amplifier.

D-C AMPLIFIERS 99

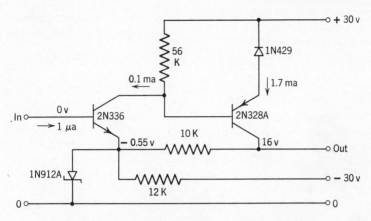

Fig. 11.2 Complementary direct-coupled d-c amplifier.

a-c application); and d-c feedback is used between the second transistors collector and the first emitter to stabilize the operating biases. It will be seen that this arrangement consists of a tandem common-emitter pair and, consequently, will have an over-all current amplification approximately equal to the product of the individual h_{fe}. There are still some deficiencies in this improved amplifier: the amount of feedback is slight, being restricted to the amount of bias variation produced by the regulation of the 1N912A as the current through it from the second stage varies, and the output voltage is still off zero for zero input, requiring some form of bucking to produce a zero reference.

A simple technique for obtaining a zero-volt output for zero input is illustrated in Fig. 11.3. It consists of an emitter-follower which would be connected to the 16-v output of Fig. 11.2. The emitter would then be at approximately 0.6 v below this voltage, or about 15.4 v. If a string of breakdown diodes, such as those shown, totaling this voltage, is inserted in the emitter lead, the low end of the diodes will be at zero volts. Making the series resistor variable, as shown, will vary the current through the diodes and, thus, the drop to some extent. Greater control may be obtained by varying one of the bias-setting resistors in the two-stage amplifier.

If only current amplification is desired, a string of n stages of cascaded emitter-followers will usually be sufficient (the Darlington pair is such a two-stage amplifier, see the preceding chapter). Such a chain will produce a progressive off-zero output voltage of n times the operating V_{BE}

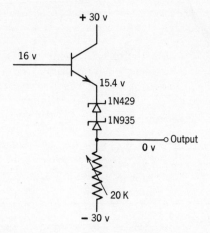

Fig. 11.3 Use of breakdown diodes to obtain zero output reference.

drop, e.g., in a three-stage amplifier, the output voltage will be about 1.8 v for zero input. To avoid this, a common-emitter stage can be used to terminate the chain, as illustrated in Fig. 11.4. This shows a chain of three emitter-followers feeding a common-emitter stage which produces the zero output when the input voltage to this stage is -1.8 v. The variable resistor in series with the emitter is used to set the output

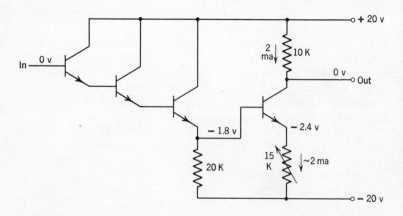

Fig. 11.4 Cascaded emitter-follower with common-emitter output stage.

at zero. A resistor of this value will produce considerable degeneration, and this may be reduced by using breakdown diodes for a major part of this voltage drop.

One other item should be mentioned. Figure 11.2 indicated that the input current is not zero, even though the voltage is. This current will depend on the input transistor, being the collector current divided by the h_{FE} of this transistor. Thus, whether the input voltage is zero or some small negative value will depend on the resistance in the input side.

Simple direct-coupled amplifiers, such as those illustrated above, are useful as long as it is not necessary to measure currents much below 1 μa. For smaller currents it becomes necessary to use differential arrangements or the various modulation techniques to be described next.

11.3 Differential Amplifiers

The major limitation to the ultimate sensitivity of a direct-coupled amplifier is the drift caused by variation of transistor biases with temperature. If this drift can be balanced by an equal drift, we can extend the range many orders of magnitude. This is the principle of the differential amplifier, which is shown in elementary form in Fig. 11.5. Two identical stages are used (for best results the two transistors are mounted in the same can, and even formed on the same chip, to achieve equal thermal conditions), and the amplifier may be used to measure the difference between the voltages on the two inputs (hence the name) or to

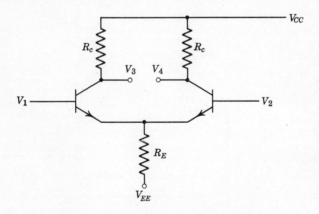

Fig. 11.5 Differential amplifier input stage.

measure one voltage, the other input being grounded. The output may be taken off one collector or the other, or differentially between the two. Thus, one differential stage may be followed by another, and the output may be converted to single-ended at a later stage, or possibly not converted, as, for example, where the output is a meter, connected differentially.

The resistor R_E may be connected to ground, in which case the two inputs must be off ground, both positive with n-p-n transistors, as shown, or it may go to a negative voltage, thus permitting the two inputs to be normally at zero voltage.

The voltage amplification of each stage of this amplifier is:

$$A_v = \frac{h_{fe}R_l}{2h_{ib}(1 + h_{fe})} \qquad (11.1)$$

where R_l is the total load resistance seen by the collector, i.e., R_c shunted by whatever input resistance the following stage presents. If the output is used differentially, i.e., with one output, V_3 going to one

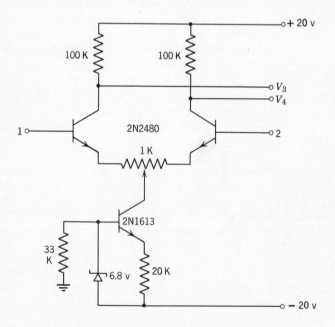

Fig. 11.6 Differential amplifier with constant-current source.

D-C AMPLIFIERS 103

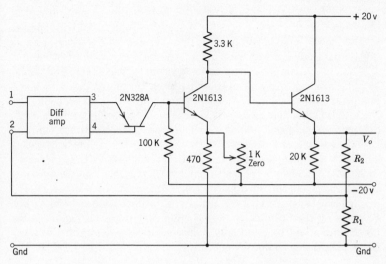

Fig. 11.7 Use of feedback to set amplifier gain.

input of a following differential amplifier, and V_4 going to the other, the over-all amplification will be twice this, as these voltages vary in opposition.

An important characteristic of differential amplifiers is the common-mode rejection. Theoretically, if the differential amplifier were perfectly balanced and the transistors were perfect devices, a common voltage imposed on both inputs simultaneously would produce no output. The ratio of the output with such simultaneously applied inputs to that obtained with only one input applied is the common-mode rejection factor. It can be calculated that this factor can theoretically be reduced to zero if the resistor R_E has the value $1/2h_{ob}$ approximately. This is an unrealistically large value and would produce excessive degeneration; however, it can be simulated by means of a third transistor, used as a constant-current source, as shown in Fig. 11.6. The 2N1613 constant-current source provides a current determined by the voltage across the breakdown diode, 6.8 v, minus the V_{BE} drop, and the resistor. Thus, for 20 kilohms, the current would be about 300 µa, or 150 per transistor. The 1-kilohm resistor between the two emitters provides a means of balancing for small inequalities between the two halves, and thus produces equal outputs. The output voltages will be about 5 v for this circuit. A second differential stage can be connected here, and its common emitter connection returned to ground through a breakdown diode, in which case the collector resistors should be changed to provide the proper voltages for the following bases. Such a dual-differential amplifier will have a common-mode rejection of 80–90 db.

A differential amplifier may be used to obtain very linear and stable amplification by applying over-all feedback to one of the differential inputs, as shown in Fig. 11.7. The differential amplifier, shown as the box in this figure, may be the single-stage amplifier of Fig. 11.6 or a dual-differential amplifier. The differential output is applied to the emitter-base connections of a p-n-p transistor, and this is followed by a cascaded common-emitter stage and an emitter-follower. The biases are adjusted such that the output voltage is normally zero when the input is zero. A portion of the output is fed back to one of the differential inputs, where it is balanced against the input signal. Thus the output voltage V_o will be given by the ratio $(R_1 + R_2)/R_1$ times the input voltage, and this amplification will be essentially independent of variation in the transistor parameters and of temperature.

11.4 Chopper-Type Amplifiers

While there are many versions of chopper-type amplifiers, all work on the basic principle of commutating an input signal to convert it to an a-c signal proportional to the input. This a-c signal is then amplified in a conventional a-c amplifier, and finally restored to a d-c output by demodulation. The key to this process is the input modulator or chopper, in this case. Figures 11.8a and 11.8b show two of the many

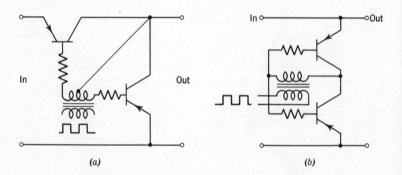

Fig. 11.8 Use of transistor choppers to convert d-c to a-c.

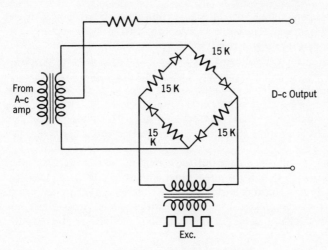

Fig. 11.9 Diode demodulator circuit.

ways in which transistors may be used to commutate the input signal. In Fig. 11.8a, one transistor switch will open the connection between input and output while the other short-circuits the output, then the reverse takes place, thus producing a push-pull effect. In Fig. 11.8b, the two switches open or close simultaneously, thus either short-circuiting the output or allowing the signal to pass.

The reason for the two transistors in Fig. 11.8b and for the reversed sequence in 11.8a is that transistors are not perfect switches. When saturated (closed condition), they become the equivalent of a small voltage in series with a small resistance. In the cutoff (open) condition, they act like small current generators, in parallel with leakage resistances. The values of these voltages, currents, and resistances depend upon whether the transistors are germanium or silicon, and whether the drive signal is applied between base and emitter or base and collector. Germanium transistors are frequently operated in the inverse condition and produce "on" voltages on the order of 0.5–1 mv, while in the "off" condition they correspond to currents on the order of 10^{-6}–10^{-7} a and to leakage resistances on the order of 10^6–10^7 ohms. Silicon transistors have voltages on the order of 30–100 mv, series resistance of 20–80 ohms in the "on" condition, 10^{-8}–10^{-10} a and 10^8–10^{10} ohms when off. Thus the choice is one of whether the higher leakage currents of the germanium transistors can be tolerated to obtain the low "on" voltage, or whether the higher voltage of the silicon transistors is acceptable to get the lower leakage. The two transistors are used in series opposition so that their "on" voltages and "off" currents will oppose each other, thus reducing the net voltage and current. Unless this balance is achieved, there will be a finite amount of a-c unbalance signal with no d-c input, the amount of this unbalance being a function of the input impedance, thus limiting the amount of this impedance. For example, chopper-type amplifiers are not practicable for use with high-impedance sources, such as ionization chambers, unless special balancing techniques are used.

As noted above, a demodulator is required to reconvert the a-c signal from the a-c amplifier to the desired d-c output. Figure 11.9 shows the circuit of a typical demodulator. The diodes are sequentially switched by the excitation signal so that the two phases of the a-c signal are connected to the same output terminal, thus producing a d-c output.

11.5 Carrier-Modulation Amplifiers

In this form of d-c amplifier, the incoming d-c signal is converted to a-c by means of some form of modulator, actuated by a high-frequency square-wave excitation (the chopper type is a form of this category of amplifier). The modulator may take the form of a diode bridge, a varactor bridge, or many other forms.

One of the most sensitive modulators uses the variation of capacitance of a diode as the basic parameter producing the modulation, as illustrated in Fig. 11.10. The two varactor diodes form a bridge with the two 330-ohm resistors and the 100-ohm balancing potentiometer. With no d-c input, the bridge is always in balance, even though the diodes are being switched from one capacitance to another by the excitation signal, since these capacitances remain equal. Thus, no a-c signal appears at the output terminals. On imposition of a d-c signal, the capacitors assume unequal values, the amount of the inequality being proportional to the d-c signal, thus there is now a net a-c signal at the amplifier input, proportional to the amplitude of the d-c input. The variable choke coil tunes the capacitance of the diodes and the a-c amplifier to approximate resonance at the excitation frequency. The remainder of the complete d-c amplifier consists of a conventional a-c amplifier, a demodulator, similar to that shown in Fig. 11.9, and occasionally over-all feedback to provide special response characteristics and additional d-c amplification, which is now at relatively high level, hence can use simpler techniques. Logarithmic response may be achieved readily with this technique by

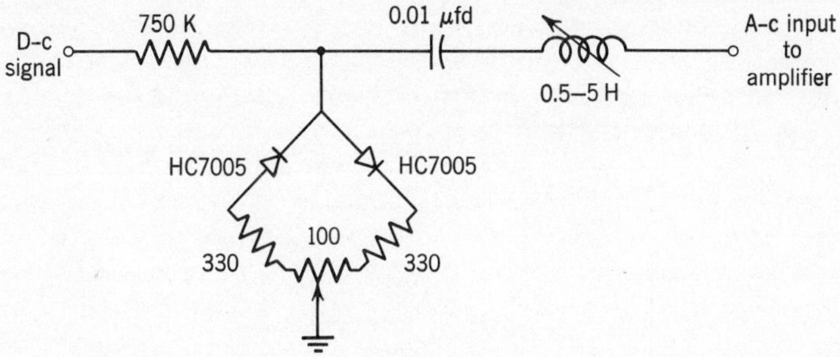

Fig. 11.10 Use of varactor diodes to modulate a d-c signal.

using a string of diodes around the whole chain, from the output of the demodulator back to the junction point of the two varactor diodes. This over-all system has been extensively described in the literature, and an amplifier has been designed having an input sensitivity below 10^{-10} a. (See Ref. 7, end of chapter.) The ultimate limit on sensitivity is dictated by the varactor diode leakage, also, in the case of logarithmic amplifiers, by that of the log-taking diodes. Using thermoelectric cooling to maintain these diodes at a very low temperature should permit attainment of lower limits on the order of 10^{-12} a by employing this technique.

11.6 Zero-Stabilized Amplifiers

One technique utilized quite extensively to extend the frequency range of chopper-type amplifiers is the combination of such an amplifier with a more conventional higher frequency amplifier, using a differential technique for the latter. In such arrangements, the input signal is applied simultaneously to one input of the differential amplifier through a-c coupling and to the input of a chopper-type amplifier. The output of the chopper amplifier is applied to the other input of the differential amplifier. This technique is illustrated in block diagram form in Fig. 11.11. The differential amplifier may take the form of that shown in Fig. 11.7, and the chopper amplifier may use one of the forms of choppers described above. For direct current, the chopper amplifier is effectively in series with one half of the differential amplifier, thus the gain is high. At higher frequencies, where the response of the chopper amplifier would

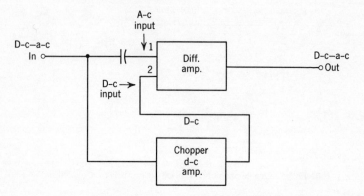

Fig. 11.11 Block diagram of zero-stabilized d-c amplifier.

normally fall off rapidly, the other half of the differential amplifier takes over. Such a stabilized amplifier typically has an input impedance on the order of 100,000 ohms, open-loop amplification greater than 100,000 at direct current, 1000 at 1 kc, and considerable utility as an operational amplifier.

11.7 Field-Effect (Unipolar) Transistors as D-C Amplifiers

Reference to Section 5.10 of Chapter 5 indicates that the field-effect transistor has characteristics very similar to those of the electron tube. Thus it should be possible to use this device in most of the circuits in which the tube has been proved advantageous. Thus, a field-effect transistor should prove comparable to an electrometer tube as a first-stage device for a d-c amplifier intended to measure currents below 10^{-10} a.

Figures 11.12a and b show two connections for the field-effect transistor similar to those used for tubes. In Fig. 11.12a the source (cathode) is connected directly to ground, in Fig. 11.12b, a resistor is employed to provide an effective negative bias for the gate (grid), in the same manner as is done with tubes.

The input impedance of a field-effect transistor is a function of drain voltage and temperature. The d-c input resistance at room temperature will normally exceed 10^9 ohms. Reference to Fig. 5.14 shows that

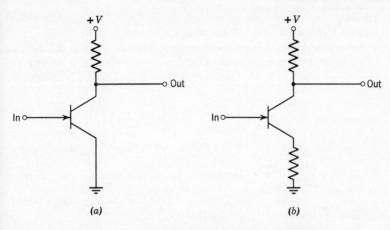

Fig. 11.12 Input stage using field-effect transistor.

variations on the order of 0.1 v on the gate produce considerable output current variations, thus this device is easily sensitive to input current changes less than 10^{-10} a, and even more sensitive devices than the one illustrated by this figure are available.

Figure 11.13 shows a hybrid combination, where the field-effect transistor is used in a manner similar to a tube cathode-follower, driving the base of a conventional transistor. This combination provides high input

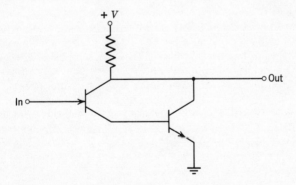

Fig. 11.13 Field-effect transistor as input stage, followed by common-emitter stage, with common load.

impedance and, at the same time, reasonably low output impedance for optimum operation of the succeeding stage.

Field-effect transistors can be operated advantageously in differential pairs to achieve the advantages of balancing drift. The succeeding stages can take any of the forms of d-c amplifiers previously described.

REFERENCES

1. Crystalonics, Inc., Low Noise Silicon Field-Effect Transistors, *Bulletin F102-362*,
2. Engineering Staff of Texas, Instruments, Inc., *Transistor Circuit Design*, McGraw-Hill, New York, 1963, pp. 124–136, 168–175, 193–196, 497–523.
3. Ekiss, J., Transistor Choppers, *Philco Corp. Application Lab Report 593A* (May, 1961).
4. Fairchild Semiconductor Fan-Out No. 109, FSP-400 Field Effect Silicon Planar Transistor (Apr. 1962).
5. Greiner, R. A., *Semiconductor Devices and Applications*, McGraw-Hill, New York, 1961, pp. 300–312.
6. Halligan, J. W., A D.C. Chopper Amplifier, *Philco Corp. Application Lab Report 731* (Oct. 31), 1961.
7. Hoge, R. R., A Sensitive Parametric Modulator for D.C. Measurements, *1960 IRE Convention Record*, Part 9, pp. 34–42.
8. Hunter, L. P. (Ed.), *Handbook of Semiconductor Electronics*, 2nd Ed., McGraw-Hill, New York, 1962, pp. 13-1–13-18.
9. Middlebrook, R. D., and A. D. Taylor, Differential Amplifier With Regulator Achieves High Stability, Low Drift, *Fairchild Semiconductor Application Data APP-45* (Dec. 1961) (reprinted from Electronics, July 28, 1961).

12

Class-A Amplifiers

12.1 Introduction

As is the case with electron tubes, we can operate transistors in any of the usual modes, class A, AB, or B. In class-A operation, the steady-state, or quiescent, operating point is set at some mid-location on the static characteristics and the signal excursion drives the output current uniformly above and below this point. In small-signal operation, the excursion is small, compared to the quiescent bias, for example, a swing of $+0.1$ ma around a central value of 1 ma. In large-signal operation, the excursion is much larger and may finally drive the operation into either cutoff or saturation, or both. If the quiescent point is properly chosen and the bias stability is adequate, the maximum excursion will produce both cutoff and saturation, on opposite peaks, thus the output is not limited by one alone.

In class-AB, the bias is offset such that peak clipping deliberately occurs at one end or the other, usually at the cutoff end. Such operation requires utilization of two transistors in push-pull to balance the distortion which would otherwise be produced, and one transistor will carry most of the load during one half of the output cycle, the other then taking over. In class-B operation, the offset is carried to the point where each transistor is biased at cutoff and only operates over one-half cycle. Class-AB and B operations of output stages will be discussed in the next chapter.

Class-A operation is commonly used for drivers which themselves feed output stages which may employ class-AB or B operation. The driver stage operates in a minor power mode, i.e., the excursion may approach large-signal conditions for the relatively low power employed in this application. This chapter will briefly discuss the basic characteristics of class-A amplifiers, ranging from small-signal operation through driver stages to power stages.

12.2 Small-Signal Class-A Operation

This mode of operation has been quite extensively covered in the preceding chapters, in particular with respect to equivalent circuits in Chapter 6, biasing in Chapter 7, and the characteristics of the single and cascaded stage in Chapters 8 and 9. Two basic arrangements are employed—transformer coupling and RC-coupling. Since the cost of transistors is now usually lower than that of transformers, and since the use of the latter results in loss of low-frequency response as well as other losses, transformers are not customarily used in transistor circuits; usually additional transistors are used to make up the loss in gain. A properly matched transformer-coupled stage will provide in excess of 50-db gain with good transistors, but two transistors will also provide this gain (refer to Table 9.1, Chapter 9), hence the choice is obvious, except where other considerations dictate the use of transformers. For example, it may be desirable to use a transformer between a driver and the output stages to provide the opposite phases required for push-pull operation and also reduce the power rating of the driver transistor.

Figure 12.1 shows the circuit of a transformer-coupled common-base stage, with the bias being obtained by the single-battery arrangement. By-pass capacitors are shown around the bias resistors. C_1 is required

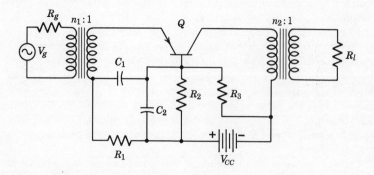

Fig. 12.1 Transformer-coupled common-base stage.

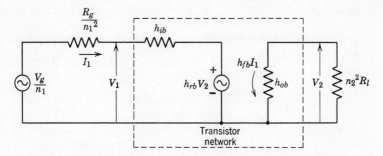

Fig. 12.2 Equivalent circuit of transformer-coupled stage.

to avoid degeneration in the emitter resistor R_1. In some cases, C_2 can be omitted, but this causes the bias resistors to be effectively in series with the load and thus reduces the efficiency and power output. Figure 12.2 shows the over-all equivalent circuit for the above arrangement, with the transistor represented by its h equivalent circuit and the generator and load resistances shown modified by the transformer turns-ratios n_1 and n_2. This network can readily be analyzed by using the network equations previously presented.

Figures 12.3 and 12.4 show the circuit of the transformer-coupled common-emitter stage and its equivalent respectively. This configuration will produce more gain, although at the expense of linearity, than the common-base configuration.

The common-collector configuration is not used in the transformer-coupled mode and will not be discussed here.

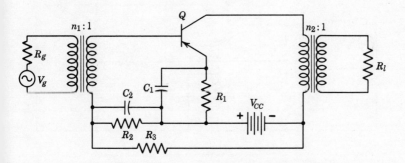

Fig. 12.3 Transformer-coupled common-emitter stage.

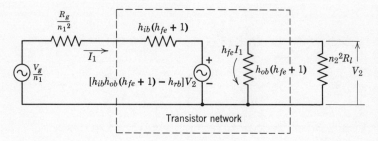

Fig. 12.4 Equivalent circuit of transformer-coupled common-emitter stage.

The common-emitter configuration is the most widely-used in RC-coupled circuits because of its higher gain. Figure 12.5 shows a typical circuit using this configuration. In this case only one by-pass capacitor is used, since the base resistor forms part of the coupling network, with R_2 and R_3 both effectively paralleling the input of the transistor. Figure 12.6 shows the equivalent circuit of this arrangement, with the transistor shown in the common-emitter h configuration using the common specification parameters. This circuit, too, can be analyzed using the previously given network equations.

The effect of the emitter impedance Z_E should be emphasized. In Chapter 8, Section 8.6, it was pointed out that introduction of such an impedance is reflected in the input circuit, multiplied approximately by the h_{fe} of the transistor; in Section 10.2, Chapter 10, this effect was

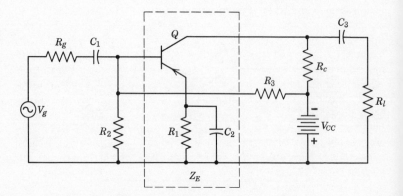

Fig. 12.5 RC-coupled common-emitter stage.

CLASS-A AMPLIFIERS   115

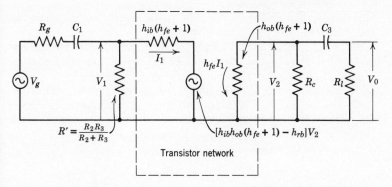

Fig. 12.6 Equivalent circuit of RC-coupled common-emitter stage. (See text for modification of transistor parameters due to nonzero value of Z_E.

analyzed by matrix methods. In Section 9.8, Chapter 9, its effect on cascaded stages was discussed, and Fig. 9.6 illustrated the effect of the by-pass capacitor on frequency response. The reader should refer to these sections for details concerning the effects of emitter impedance.

12.3 Linearity Considerations in Class-A Amplifiers

Since the input circuit of a transistor is extremely nonlinear, it is to be expected that the linearity of relationship between input signal and output signal will depend to a marked degree on the character of the input signal, i.e., whether it is a sinusoidal input current, for example, or a sinusoidal input voltage, and which transistor configuration is used. Since the common-base configuration has the most linear set of output characteristics, it is to be expected that the relationship between an input current applied to the emitter and the collector current would be quite linear, and this is the case. On the other hand, the relationship between a voltage applied between emitter and base and collector current should have essentially the same characteristic as the input family, hence should be very nonlinear. Figure 12.7, which shows collector current versus input signal, illustrates this. The curve of I_C versus I_E is very linear, that for I_C versus V_{EB} is very nonlinear. The third curve, I_C versus V_G through 100 ohms, shows that a reasonable compromise can be achieved by using a moderate amount of generator

116   TRANSISTOR APPLICATIONS

impedance. The greater this impedance the more linear the curve, but the greater the loss.

Figure 12.8 shows similar curves for the common-emitter configuration. Here we have two effects which compensate each other to some degree. In addition to the input circuit nonlinearity we have also the effect of the crowding of the characteristics at high current. Thus, the curve of I_C versus I_B is not linear but bends at the upper end, whereas the curve of I_C versus V_{BE} bends at the lower end. The implication is that there should be an optimum value of source impedance, and this is the case, as shown by the other two curves. A generator impedance

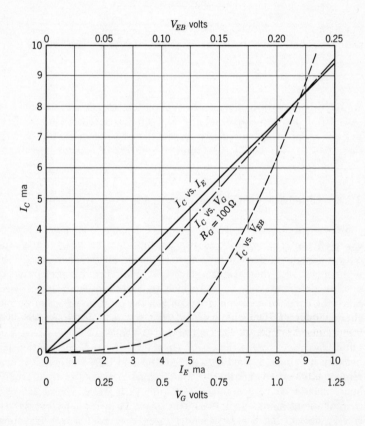

Fig. 12.7   Output current versus input current and voltage, common-base configuration.

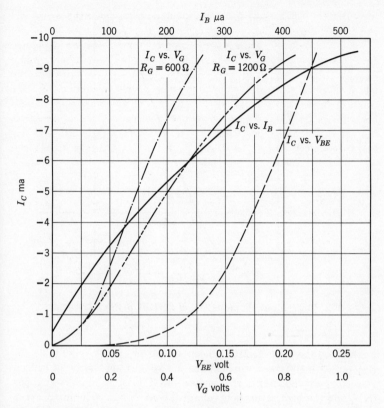

Fig. 12.8 Output current versus input current and voltage, common-emitter configuration.

on the order of 1000 ohms would evidently provide about the best compromise between the two sources of distortion.

12.4 High-Power Class-A Amplifiers

Although transistor power amplifiers usually use the class-AB or B modes, because of the greater efficiency and lower stand-by power requirements, there are still occasions where high-power class-A amplifiers are used. Figures 12.9 and 12.10 show the input and collector

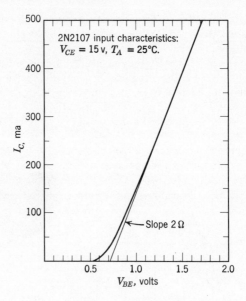

Fig. 12.9 Input characteristic of 2N2107 power transistor.

families for the type-2N2107 silicon transistor, which is suitable for moderately high-power applications. Two characteristics of silicon transistors are apparent; the higher cutoff bias, in this case about 0.5 v, and the high saturation voltage, about 2–3 v. The input curve approaches a slope of about 2 ohms, which is a measure of the high-current value of h_{ib}. h_{FE} is around 50, and the corresponding value of h_{ie} is therefore about 100 ohms.

A load line of 100 ohms has been drawn on Fig. 12.10 to illustrate typical operation. By employing this load line, we can deduce several characteristics of the class-A amplifier using this transistor in this mode. The maximum swing between cutoff and saturation is 40 v and 400 ma. The rms power output is given by the relationship

$$P_o = \frac{(V_{\max} - V_{\min})(I_{\max} - I_{\min})}{8} \qquad (12.1)$$

$$= 40 \times 0.4 \div 8 = 2.0 \text{ w}$$

To achieve this power output, the quiescent bias should be at mid-swing, or at about 24 v, 210 ma, for a quiescent power dissipation of 5 w. Thus, the efficiency is ⅖ or 40%. This indicates one of the disadvantages of class-A operation at high power, as considerably better efficiency can be obtained with class B. Furthermore, to obtain even this efficiency, the quiescent operating point cannot move, hence the requirements on bias stiffness would be severe, requiring dissipation in the bias resistors comparable to that in the transistor, and further reducing the efficiency. Thus, if this amplifier were required to operate over any appreciable temperature range, it would be limited to an output of about 1 w, with an efficiency of about 15–20%.

The input power, and hence power gain, can also be obtained from these curves. The base-current excursion corresponding to the above 400-ma collector-current swing is about 10 ma, and the base-voltage swing is about 1 v. Using the same calculation as above, we obtain for the input power,

$$P_i = 0.010 \times 1 \div 8 = 0.00125 \text{ w}$$

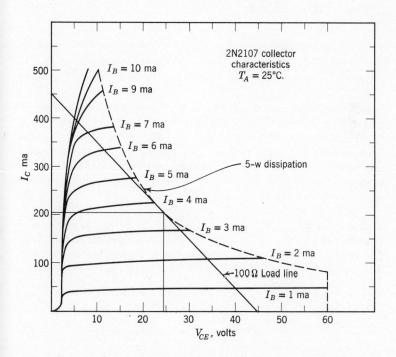

Fig. 12.10 Collector characteristics of 2N2107 transistor.

Power gain is the ratio

$$G = 2.0 \div 0.00125 = 1600 \text{ or } 32 \text{ db}$$

Thus, the power gain is quite high, and such a transistor can be used very effectively to drive higher power stages.

REFERENCES

1. Hunter, L. P., *Handbook of Semiconductor Electronics*, McGraw-Hill, New York, 1962, pp. 11-39–11-46.
2. Shea, R. F., *Transistor Audio Amplifiers*, Wiley, New York, 1955, pp. 124–162.
3. General Electric Co., *Transistor Manual*, Sixth ed., pp. 113–124.
4. Motorola Semiconductor Products Division, Inc., *Motorola Power Transistor Handbook*, First ed., pp. 45–71.
5. Texas Instruments Staff, *Transistor Circuit Design*, McGraw-Hill, New York, 1963, pp. 206–213.

13

Class-*B* Amplifiers

31.1 Introduction

In Chapter 12, class-*A* amplifiers were described for both small-signal and large-signal applications. Class-*B* amplifiers are only used for relatively large signals, although, occasionally, this mode of operation may be used for driving higher power stages. The major advantages of using class *B* are the very low stand-by power requirements, the relatively high efficiency, and thus high power output from relatively low-power transistors. The disadvantages are higher distortion, possibility of crossover distortion, and possibly more stringent bias stability requirements.

13.2 Basic Considerations of Class-*B* Operation

In this mode of operation, the transistor is biased normally to cutoff and the input signal drives the output up toward saturation. Push-pull operation is required, with each half taking the load on alternate half-cycles. Figure 13.1 shows how the collector current of one transistor varies from a value approximately equal to I_{CBO} during the nonconducting portion of the cycle, up to a peak value of I_C'. Figure 13.2 shows the variation of collector voltage during the cycle. During the conducting portion, the collector voltage drops to a low value, approaching zero (actually the saturation voltage). When this transistor is cut off, the voltage rises to approximately twice the supply voltage by transformer action from the other half. The instantaneous dissipation,

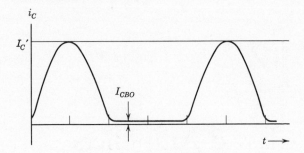

Fig. 13.1 Class-B output current.

therefore, has two peaks, with low-dissipation areas between, as shown in Fig. 13.3.

Operation of class-B amplifiers can be analyzed graphically by means of the static characteristics. Referring to Fig. 12.10, class-B operation implies that the quiescent operating point would be located at the lower extremity of the load line, in this example at $V_{CE} = 45$ v, $I_C = 0$ (assuming I_{CBO} is negligible). The stand-by dissipation is therefore zero. A peak collector-current swing of about 400 ma can be used, with corresponding peak collector-voltage swing of about 40 v. Using two stages, this produces a peak-to-peak current swing of 0.8 a, voltage swing of 80 v, or an output power of

$$P_o = 0.8 \times 80 \div 8 = 8 \text{ w}$$

or 4 w per transistor. As an approximation we can say that the output

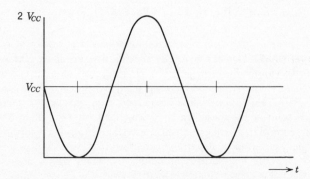

Fig. 13.2 Class-B collector voltage.

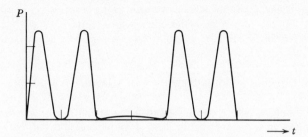

Fig. 13.3 Instantaneous collector power.

power of a pair of transistors, operated in push-pull class B, is

$$P_o \simeq \frac{V_{CC}(I_C' - I_{CBO})}{2} \tag{13.1}$$

or, if I_{CBO} is negligible,

$$P_o \simeq \frac{V_{CC}I_C'}{2} \tag{13.2}$$

In this example this would give us $45 \times 0.4 \div 2$ or 9 w. The loss of 1 w is due to the inability to drive the transistor to the zero-voltage axis. In actual practice, the above full power would not be practicable, since distortion would be excessive at the peaks because of crowding of the characteristics, and, as with class A, excessive power dissipation would be necessary to hold the bias point firm against temperature drift.

The average value of the current shown in Fig. 13.1 is given by

$$I_{C\,\text{av}} = \frac{I_C' + I_{CBO}(\pi - 1)}{\pi} \tag{13.3}$$

The power supplied by the battery per transistor is this current multiplied by the supply V_{CC}, or

$$P_{\text{d-c}} = 2V_{CC}\frac{I_C' + I_{CBO}(\pi - 1)}{\pi} \tag{13.4}$$

per pair of transistors. The efficiency is therefore the total power output, as given by eq. 13.1, divided by total power supplied, from eq. 13.4, or

$$\eta = \frac{\pi(I_C' - I_{CBO})}{4[I_C' + I_{CBO}(\pi - 1)]} \tag{13.5}$$

13.3 Linearity Considerations in Class-B Amplifiers

If I_{CBO} is negligible, the efficiency at full swing becomes $\pi/4$ or approximately 78%. Class-B amplifier efficiencies around 75% are relatively easy to obtain.

Curves of output versus input can be constructed for class-B amplifiers in the same manner as for class-A amplifiers, and Fig. 13.4 shows such a set of curves for the 2N2107 transistor, for which the static

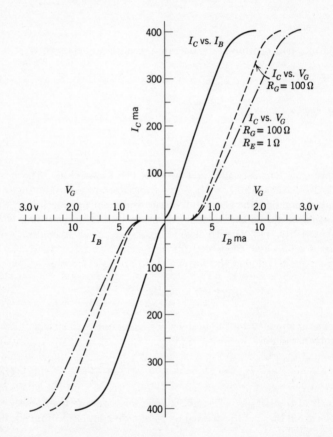

Fig. 13.4 Output current versus input current and voltage, 2N2107 in push-pull class-B operation.

CLASS-B AMPLIFIERS   125

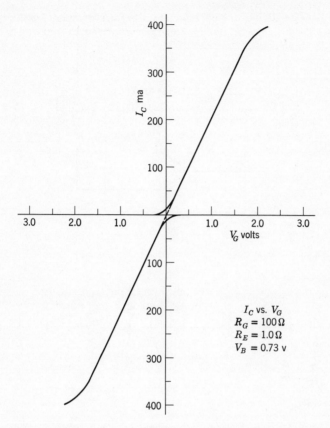

Fig. 13.5 Application of forward bias to eliminate cross-over distortion.

characteristics were presented in Chapter 12. These curves are obtained by taking points off the 100-ohm load line of Fig. 12.10 and obtaining corresponding values of input voltage from Fig. 12.9, then constructing mirror images to produce the push-pull effect. It will be noted that even the I_C versus I_B curves are displaced slightly and are not direct continuations of each other, while the I_C versus V_G curves show great discontinuity. This discontinuity is the cause of the so-called crossover distortion, wherein the output current would have the shape of portions of sinusoids instead of full sinusoids. One particularly objectionable

126 TRANSISTOR APPLICATIONS

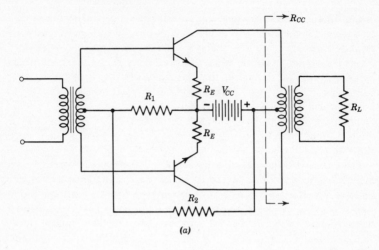

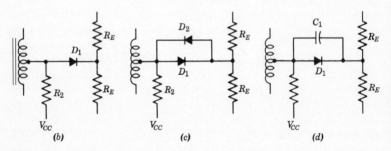

Fig. 13.6 Class-B biasing techniques. (a) Basic class-B schematic. (b-d) Alternative biasing arrangements.

feature of this form of distortion is that it occurs at low levels, hence is extremely undesirable.

13.4 Elimination of Crossover Distortion in Class-B Amplifiers

As indicated above, crossover distortion is produced by nonalignment of the two halves of the composite push-pull characteristic; therefore the cure should be some means of realigning these two halves. Inspection of Fig. 13.4 indicates that this may be accomplished by applying a small amount of forward bias to the amplifier, so that the characteristics are displaced toward the axis. Figure 13.5 illustrates this technique. These curves are the same as the dotted curves of Fig. 13.4, displaced by an amount 0.73 v, so that the linear portions are now extensions of each other. Under this condition, the transition from one transistor to the other will be smooth and there will be little evidence of crossover distortion. In effect, we have modified the class-B stage to be a class-AB stage, although the forward bias is still quite low and operation is much closer to class B than to A.

The methods employed to obtain this forward bias are illustrated in Fig. 13.6. In Fig. 13.6a two small resistors are inserted in the emitter leads to improve bias stability in the same manner as for small-signal stages, and a portion of the battery voltage is applied to the bases through the divider R_1, R_2. The resistors R_E are usually quite small, on the order of a few ohms, with the drop across them being about 0.5–1 v. As indicated above, the drop across R_1 is in the tenths of volts, just enough to eliminate the crossover misalignment.

One additional problem involves the variation of V_{BE} with temperature. As noted in a previous chapter, this can be on the order of 1–2 mv/°C, hence can produce quite a shift if a wide temperature range is to be accommodated. Figures 13.6b, c, and d illustrate methods of varying the forward bias on the transistors in about the same manner as the shift in emitter-base voltage. All involve obtaining the forward bias from a current through a diode, which will have essentially the same temperature variation as the transistor. In Fig. 13.6a, a diode simply replaces R_1. In Fig. 13.6b, two diodes are used, the purpose of D_2 being to absorb the reverse base current on extreme swings, and in Fig. 13.6d this is accomplished by the capacitor C_1. Normally capacitors are not used to by-pass the resistors of a class-B stage since the current through these resistors is unidirectional, and the capacitor will charge and change the operating point toward cutoff or actually beyond it. A capacitor can be used across a diode, however, since the latter is essentially a constant-voltage device.

13.5 Complementary Class-B Amplifiers

In this and the next section, variations of the basic class-B amplifier will be described, which possess rather unique features. One of these is the complementary amplifier, which utilizes opposite types of transistors, hence is possible only with transistors. Figure 13.7 illustrates this

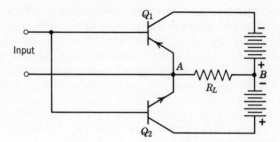

Fig. 13.7 Complementary class-B output stage.

technique. A p-n-p and a n-p-n transistor are used with their inputs tied together, and the load is connected between the two emitters and the center tap of the supply. When the input is positive with respect to the low end of the input transistor, Q_2 conducts and supplies output current to the load R_L. When the input reverses, Q_1 takes over, thus the operation is similar to that described previously. One disadvantage of this arrangement is that either one side of the input or the center tap of the battery can be on ground, but not both. If point A is grounded, the batteries must be off ground and cannot be common to other circuits. If point B is grounded, permitting common batteries, the input must be off ground, e.g., by using a transformer. No bias resistors are required on the input since each transistor acts as a base return for the other; in fact, the input can employ a blocking capacitor without risking thermal runaway.

13.6 Low-Output-Impedance Direct-Drive Amplifier*

In the amplifier to be described, advantage is taken of the fact that transistors permit the attainment of very low output impedance, so that they can be used to drive a loudspeaker voice coil directly without interposition of an output transformer with its attendant losses. The principle is similar to that described in the foregoing, except that complementary transistors are used to drive similar-type output transistors. Several other unique features are incorporated in this amplifier, making it an excellent example of the potential applications of transistors in this manner.

* From Ref. 4, end of chapter.

Figure 13.8 gives the schematic of a 10-w amplifier employing this principle. Transistors Q_2 and Q_4 form one Darlington pair, Q_3 and Q_5 another, with each output transistor supplying the output alternately. Use of the p-n-p transistor Q_2 and the n-p-n transistor Q_3 provides the phase difference for the output transistors, and permits operation from the same input transistor Q_1. Q_4 and Q_5 have a small forward bias of 10–20 ma to minimize crossover distortion and to also operate the output transistors in a more favorable beta region, this bias being supplied by the drop across the two 390-ohm resistors. Q_2 and Q_3 are biased at about 1 ma by the two 1N91 diodes. These diodes also provide temperature compensation, as described previously. The 47-ohm resistor in the emitter of Q_3 further improves stabilization and also provides feedback to decrease distortion.

Q_1 is a class-A driver with an emitter current of about 3 ma. Nega-

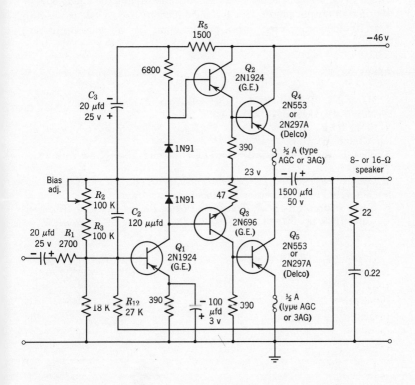

Fig. 13.8 10-watt amplifier employing direct output drive.

tive feedback to its base is provided by means of R_2 and R_3 and also through R_{12}. Positive feedback is also provided by means of C_3 and R_5 to compensate for the unsymmetrical output circuit and to permit the positive peak signal swing to approach the amplitude of the negative peak. The net amount of negative feedback over-all is approximately 14 db. The value of C_2 was determined by optimizing the square-wave response.

This amplifier has an output impedance of about 1 ohm, providing good speaker damping, low distortion, and good bandwidth. The power response at 1 w is flat from 30 cycles to 15 kc and is down 3 db at 50 kc. At this level, the total harmonic and intermodulation distortion are both less than 1%. At 7 w, the intermodulation distortion is less than 2½% and the total harmonic distortion is less than 1% measured at 50 cycles, 1 kc, and 10 kc. The performance is about the same for either 8- or 16-ohm loads. The amplifier is capable of about 8 w of continuous output power with 1-v rms input, or 10-w music power into 8 or 10 ohms when used with a power supply having good regulation.

REFERENCES

1. Hunter, L. P., *Handbook of Semiconductor Electronics*, McGraw-Hill, New York, 1962, pp. 11-46–11-62.
2. Shea, R. F., *Transistor Audio Amplifiers*, Wiley, New York, 1955, pp. 163–188.
3. Shea, R. F. (Ed.), *Transistor Circuit Engineering*, Wiley, New York, 1957, pp. 113–137.
4. General Electric Co., *Transistor Manual*, Sixth ed., pp. 113–140.
5. Motorola Power Transistor Handbook, pp. 45–100.
6. Texas Instruments Staff, *Transistor Circuit Design*, McGraw-Hill, New York, 1963, pp. 220–232.

14

High-Frequency Parameters

14.1 Introduction

In Chapter 6 we presented the equations for the h parameters in terms of four basic parameters of the ideal transistor, plus parasitic base resistance and collector capacitance. The equations for the real transistor parameters were then given in terms of these basic units, the cutoff frequency, and the operating frequency. We thus obtained equations from which we can obtain the complex values of the transistor parameters at any frequency. In this chapter we will review these equations and indicate how the parameters vary as the cutoff frequency is approached. This will provide a basis to the study of high-frequency circuits in subsequent chapters.

14.2 The Basic Parameters

The basic parameters were given in Chapter 6 and are repeated here for convenience.

$$r_e = \frac{kT}{qI_E} \simeq \frac{25.6}{I_E} \quad \text{(at room temperature, } I_E \text{ in ma)} \quad (14.1)$$

$$r_b' \simeq \frac{(h_{ib})_0 - r_e}{1 + (h_{fb})_0} \quad (14.2)$$

$$\simeq [(h_{ib})_0 - r_e][1 + (h_{fe})_0] \quad (14.3)$$

$$\alpha_{b0} = -(h_{fb})_0 = \frac{(h_{fe})_0}{1 + (h_{fe})_0} \tag{14.4}$$

$$g_c \simeq (h_{ob})_0 \tag{14.5}$$

$$\mu_0 \simeq (h_{rb})_0 - r_b'g_c \tag{14.6}$$

In the above equations the ( )$_0$ indicates the low-frequency specification value of the parameter. It should be borne in mind that the above equation for r_b' is only an approximation, and a poor one at best, being the difference between two nearly equal quantities; however, if the specification does not supply a value for r_b', it may be the only way to approximate its value.

14.3 Parameters of the Actual Transistor at High Frequencies

Equations 6.26–6.29 gave the values of the common-base h parameters as functions of frequency, expressed in terms of the above basic parameters. Equations 6.30–6.32 gave simplified versions of these equations which are adequate for the great majority of needs. These are repeated here for convenience.

$$h_{ib} = \frac{1}{1 + j\omega/\omega_{ab}} \left[r_\epsilon + r_b'\left(1 - \alpha_{b0} + \frac{j\omega}{\omega_{ab}}\right) \right] \tag{14.7}$$

$$h_{rb} = (h_{rb})_0 + j\omega r_b' C_c \tag{14.8}$$

$$h_{fb} = \frac{(h_{fb})_0}{1 + j\omega/\omega_{ab}} \tag{14.9}$$

$$h_{ob} = (h_{ob})_0 + j\omega C_c \tag{14.10}$$

It is also desirable to know the values of the common-emitter parameters directly, in terms of the low-frequency spec values. The two most widely used at high frequencies are h_{ie} and h_{fe}. These are given by the following:

$$h_{ie} = \frac{(h_{ib})_0 + j\omega r_b'/\omega_{ab}}{1 + j\omega/\omega_{ab} + (h_{fb})_0} \tag{14.11}$$

$$= \frac{(h_{ib})_0 + j\omega r_b'/\omega_{ab}}{1/[1 + (h_{fe})_0 + j\omega/\omega_{ab}]} \tag{14.12}$$

$$h_{fe} = \frac{(h_{fe})_0}{1 + j\omega[1 + (h_{fe})_0]/\omega_{ab}} \tag{14.13}$$

HIGH-FREQUENCY PARAMETERS

At frequencies approaching the cutoff frequency the above equations can be even further simplified, as below:

$$h_{ib} \simeq \frac{r_\epsilon + j\omega r_b'/\omega_{ab}}{1 + j\omega/\omega_{ab}} \tag{14.14}$$

$$h_{rb} \simeq j\omega r_b' C_c \tag{14.15}$$

$$h_{fb} \simeq \frac{(h_{fb})_0}{1 + j\omega/\omega_{ab}} \tag{14.16}$$

$$h_{ob} \simeq j\omega C_c \tag{14.17}$$

$$h_{ie} = r_b' - \frac{j\omega_{ab} r_\epsilon}{\omega} \tag{14.18}$$

$$h_{re} = \omega_{ab} r_\epsilon C_c \tag{14.19}$$

$$h_{fe} = \frac{-j\omega_{ab}(h_{fe})_0}{\omega[1 + (h_{fe})_0]} \tag{14.20}$$

$$h_{oe} = \omega_{ab} C_c \left(1 + \frac{j\omega}{\omega_{ab}}\right) \tag{14.21}$$

It is interesting to note how the parameters change as frequency increases. For example, consider a transistor having the following low-frequency parameters: (at $I_E = 1$ ma)

$$h_{ib} = 29 \quad h_{rb} = 4 \times 10^{-4} \quad h_{fb} = -0.98 \quad h_{ob} = 0.6 \times 10^{-6}$$
$$f_{ab} = 1.2 \text{ Mc} \quad C_c = 35 \times 10^{-12}$$

We calculate the following basic parameters:

$$r_\epsilon = 25.6 \text{ ohms} \quad r_b' = 170 \text{ ohms} \quad \mu_0 = 3 \times 10^{-4} \quad \alpha_{b0} = 0.98$$
$$g_c = 0.6 \times 10^{-6} \text{ mho}$$

From eqs. 14.7–14.13 we calculate the parameters at a frequency of 1 Mc:

$$h_{ie} = 170.7 - j30.7$$
$$h_{re} = (67 + j1.6)10^{-4}$$
$$h_{fe} = 0.022 - j1.18$$
$$h_{oe} = (262 + j227)10^{-6}$$

Using the simple equations 14.18–14.21, we get:

$$h_{ie} = 170 - j30.7 \qquad h_{fe} = 0 - j1.18$$
$$h_{re} = 67.4 \times 10^{-4} \qquad h_{oe} = (262 + j219)10^{-6}$$

Several observations are possible from the above results. First, it is evident that the accuracy afforded by the simplified equations is more than adequate at this frequency. Second, the pronounced change in the parameters with frequency should be noted. For example, h_{ie} changes from a low-frequency value of 1420 ohms to a value of 173 ohms at 1 Mc, and has a considerable capacitive reactance. The current multiplication factor h_{fe} has also changed greatly, from a value of 49 at low frequency to only 1.18 at 1 Mc, and is completely reactive at that frequency. By comparison, the common-base amplification can also be computed and turns out to have a magnitude of 0.74 at 1 Mc, thus the great superiority of the common-emitter stage is rapidly reduced as the cutoff frequency is approached. Other considerations, e.g., stability, still make the common-emitter the preferential configuration at high frequencies.

14.4 The High-Frequency Equivalent Circuit

As the frequency approaches the cutoff frequency, the more complex equivalent circuit of Chapter 6 may be simplified to the form shown in Fig. 14.1.

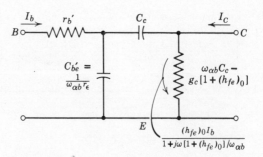

Fig. 14.1 High-frequency equivalent circuit.

14.5 Typical Parameter Variation with Frequency

The following curves, which show typical variations of transistor parameters with frequency, are reproduced from the Texas Instruments RF Seminar Papers. The y parameters are used in these illustrations, primarily because these parameters are relatively easy to measure at frequencies in the hundreds of megacycles.

The 2N2415 transistor, for which these curves are drawn, is a germanium mesa transistor. The specification sheet gives a low-frequency value of h_{fe} of 45, dropping to 2.8 at 200 Mc. The specified value of the output capacitance, C_{ob} is 1.2 pf, the base-spreading resistance-collector capacitance product, $r_b'C_c$, is given as 3.5×10^{-12} second, and the transition frequency f_T (the frequency at which h_{fe} drops to unity) is given as 560 Mc.

All the curves were measured at values of $V_{CE} = -6$ v, $I_C = -2$ ma, at 25°C.

Figure 14.2 shows the variation of the real and imaginary parts of the input admittance y_{ie} over the frequency range of 50–900 Mc. Both curves show maxima at the upper-frequency ends.

Figure 14.3 shows the variation of the real and imaginary parts of the

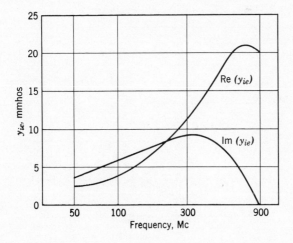

Fig. 14.2 Variation of y_{ie} with frequency, 2N2415 transistor. (From Texas Instruments RF Seminar Papers.)

feedback admittance y_{re}. Both are negative values, and both increase with frequency.

Figure 14.4 shows the variation of the forward transfer admittance y_{fe}. The real part decreases uniformly, the imaginary part is negative and peaks at 300 Mc.

Figure 14.5 shows the variation of y_{oe}, the output admittance. Both parts increase with frequency.

If it is desired to obtain the high-frequency h parameters, they may be readily calculated from these y parameters, using the transformation equations:

$$h_{ie} = \frac{1}{y_{ie}} \qquad h_{re} = -\frac{y_{re}}{y_{ie}}$$

$$h_{fe} = \frac{y_{fe}}{y_{ie}} \qquad h_{oe} = \frac{\Delta^y}{y_{ie}}$$

Using the above interrelationships, the values of h_{ie} and h_{fe} were calculated from the curves of Figs. 14.2 and 14.4. Figure 14.6 shows the thus calculated values of the real and imaginary parts of the input impedance h_{ie}. Of particular note is the very low value of the real portion, compared to the low-frequency value, which would be on the order

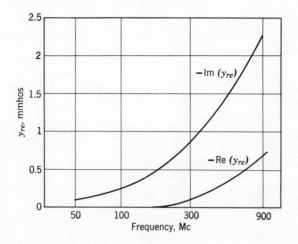

Fig. 14.3 Variation of y_{re} with frequency, 2N2415 transistor. (From Texas Instruments RF Seminar Papers.)

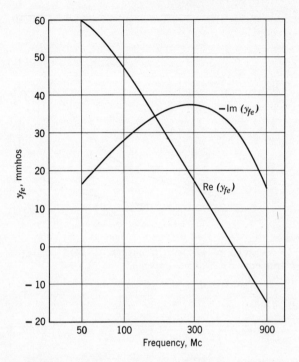

Fig. 14.4 Variation of y_{fe} with frequency, 2N2415 transistor. (From Texas Instruments RF Seminar Papers.)

of 1000 ohms. The reactive component is capacitive, except above about 900 Mc. This crossover will be obviously a function of the lead inductance.

Figure 14.7 shows the variation of the magnitude of h_{fe} over this frequency range. The transition frequency, f_T is seen to be at about 900 Mc, indicating that the specification value of 560 Mc is on the conservative side. If this curve were carried out to lower frequencies, it would level off at the low-frequency value of 45 at about 1 Mc, or possibly higher, indicating the excellent potential performance of this transistor at the usual radio frequencies.

It is also interesting to calculate the high-frequency value of h_{fe} from its low-frequency value and the cutoff frequency from the relationship

$$h_{fe} \simeq \frac{(h_{fe})_0}{1 + j\omega[1 + (h_{fe})_0]/\omega_{ab}}$$

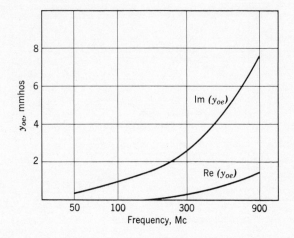

Fig. 14.5 Variation of y_{oe} with frequency, 2N2415 transistor. (From Texas Instruments RF Seminar Papers.)

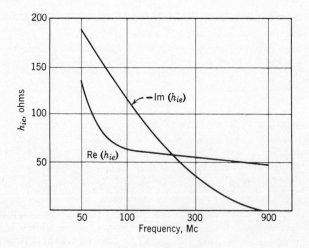

Fig. 14.6 Variation of h_{ie} with frequency, 2N2415 transistor.

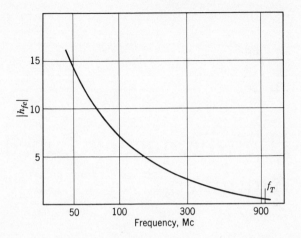

Fig. 14.7 Variation of the magnitude of h_{fe} with frequency, 2N2415 transistor.

Assuming that the cutoff frequency is approximately equal to the transition frequency, f_T, and using 900 Mc for this value, we calculate h_{fe} at 300 Mc to be $2.93 / -86.3°$, as compared to the value of $2.71 / -98.8°$ obtained from the y parameters. Considering the uncertainty of the cutoff frequency and the fact that the above equation is only an approximation, the agreement is quite good.

REFERENCES

1. Hunter, L. P., *Handbook of Semiconductor Electronics*, McGraw-Hill, New York, 1962, pp. 12-2–12-12–12-29.
2. Martinengo, R., High Frequency Parameters of PNP Fusion Alloy Transistors, *Raytheon Technical Information Bulletin TIS-111-T* (Sept. 1958).
3. Shea, R. F. (Ed.), *Transistor Circuit Engineering*, Wiley, New York, 1957, pp. 34–47.
4. Philco Semiconductor Data Sheet, Supplement 15 (Sept. 1961).
5. Valdes, L. B., *The Physical Theory of Transistors*, McGraw-Hill, New York, 1961, pp. 323–334.
6. Texas Instruments RF Seminar Papers, Texas Instruments Inc., Dallas, Texas, 1963.

15

High-Frequency Tuned Amplifiers

15.1 Introduction

In the preceding chapter, the high-frequency parameters of the transistor were discussed, and the fact that they are extremely variable with frequency was emphasized. In this chapter we will use these high-frequency parameters in the design of high-frequency tuned amplifiers, and in the next chapter the subject of wide-band amplifiers will be discussed.

15.2 The Concept of Normalized Bandwidth

Consider the elementary single-tuned circuit of Fig. 15.1, consisting of the inductance L, capacitance C, and conductance g, which is the sum of the conductances of the other two elements. The network is supplied by a constant-current source I, and an output voltage V appears across it. The input impedance of this shunt combination is:

$$Z = \frac{j\omega L}{1 - \omega^2 LC + j\omega g L} \tag{15.1}$$

The Q factor of this combination may be expressed in terms of the inductance:

$$Q = \frac{1}{\omega g L} \tag{15.2}$$

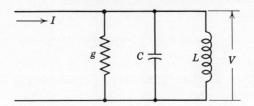

Fig. 15.1 Components of parallel-tuned circuit.

We can also express the impedance in terms of the Q:

$$Z = \frac{j\omega L}{1 - \omega^2 LC + j/Q} \tag{15.3}$$

At resonance, $\omega_0^2 LC = 1$ and Z becomes $1/g$.
Off resonance we have:

$$\frac{V}{V_0} = \frac{j\omega gL}{1 - \omega^2 LC + j\omega gL} = \frac{j/Q}{1 - \omega^2 LC + j/Q} \tag{15.4}$$

Let us now designate n as a number such that B_n is the bandwidth for a power output $1/(n+1)$ of that at resonance (e.g., B_1 is the half-power or 3-db bandwidth, B_3 is the 6-db bandwidth, etc.). By this definition,

$$n = \frac{P_0}{P} - 1 = \left|\frac{V_0}{V}\right|^2 - 1 \tag{15.5}$$

From the previous equation for V/V_0 we obtain,

$$\left|\frac{V_0}{V}\right|^2 = Q^2(1 - \omega^2 LC)^2 + 1 \tag{15.6}$$

whence

$$n = Q^2(1 - \omega^2 LC)^2 \tag{15.7}$$

and

$$1 - \omega^2 LC = \frac{\sqrt{n}}{Q} \tag{15.8}$$

This gives us the two frequencies defining the bandwidth:

$$\omega_{n1} = \sqrt{\frac{1 + \sqrt{n}/Q}{LC}} \quad \text{and} \quad \omega_{n2} = \sqrt{\frac{1 - \sqrt{n}/Q}{LC}} \tag{15.9}$$

By definition,

$$B_n = \frac{\omega_{n1} - \omega_{n2}}{2\pi} \tag{15.10}$$

Substituting eq. 15.9 into eq. 15.10 and assuming that $\sqrt{n/Q} \ll 1$, eq. 15.10 approximates to:

$$B_n = \frac{\sqrt{n}}{2\pi Q \sqrt{LC}} \tag{15.11}$$

Normalizing by dividing by the resonant frequency f_0, we obtain the *normalized bandwidth*

$$B_n' = \frac{\sqrt{n}}{Q} \tag{15.12}$$

15.3 Source and Load Coupled by a Single-Tuned Circuit*

Figure 15.2 shows the equivalent circuit of a single-tuned circuit, such as that of Fig. 15.1, used to couple a source having an output conductance g_{o1} and capacitance C_{o1} to a load having input conductance g_{i2}' and capacitance C_{i2}'. The primes indicate that these values are trans-

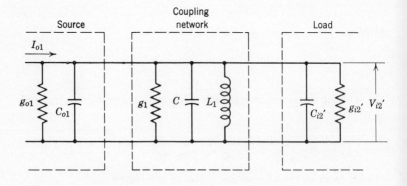

Fig. 15.2 Addition of output and input loading to tuned circuit.

* See Ref. 4, end of chapter, pp. 160–174.

HIGH-FREQUENCY TUNED AMPLIFIERS 143

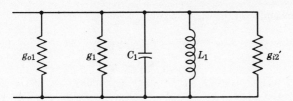

Fig. 15.3 Composite tuned circuit with loads.

formed from the actual input values of the following stage. The source supplies a constant current I_{o1} and there is an output V_{i2}'. Again, the actual input voltage to the following stage, V_{i2} will be less than this by the transformer ratio.

The three capacitances can be combined, as shown in Fig. 15.3, to form one capacitance C_1. The characteristics of this simplified circuit can now be determined.

The power available from the source is:

$$P_1 = \frac{|I_{o1}|^2}{4g_{o1}} \tag{15.13}$$

At resonance, the power delivered to the load is

$$P_2 = \frac{|I_{o1}|^2 g_{i2}'}{(g_{o1} + g_1 + g_{i2}')^2} \tag{15.14}$$

The power transfer efficiency is the ratio of output power to available power:

$$\eta = \frac{P_2}{P_1} = \frac{4g_{o1}g_{i2}'}{(g_{o1} + g_1 + g_{i2}')^2} \tag{15.15}$$

The Q of the tuned circuit has been reduced by the addition of g_{o1} and g_{i2} to

$$Q = \frac{g_1 Q_0}{g_{o1} + g_1 + g_{i2}'} \tag{15.16}$$

Substituting this in the equation for normalized bandwidth:

$$B_n' = \frac{\sqrt{n}}{Q} = \frac{\sqrt{n}\,(g_{o1} + g_1 + g_{i2}')}{g_1 Q_0} \tag{15.17}$$

Equation 15.17 can be solved to obtain g_1:

$$g_1 = \frac{\sqrt{n}\,(g_{o1} + g_{i2}')}{Q_0 B_n' - \sqrt{n}} \tag{15.18}$$

Remembering that $g_1 Q_0 = 1/\omega L_1$, we can obtain the required value of tuning inductance, in terms of the specified bandwidth, loading conductances, and an estimate of the available unloaded Q:

$$L = \frac{B_n' - \sqrt{n}/Q_0}{\omega \sqrt{n}\,(g_{o1} + g_{i2}')} \tag{15.19}$$

Substituting eq. 15.18 into eq. 15.15, we obtain:

$$\eta = \frac{4 g_{o1} g_{i2}' (Q_0 B_n' - \sqrt{n})^2}{[Q_0 B_n' (g_{o1} + g_{i2}')]^2} \tag{15.20}$$

If g_{i2} is matched to g_{o1} we obtain maximum transfer efficiency:

$$\eta_{\max} = \left(1 - \frac{\sqrt{n}}{Q_0 B_n'}\right)^2 \tag{15.21}$$

This matching will require a transformer having a turns ratio m given by $m = \sqrt{g_{i2}/g_{o1}}$, where g_{i2} is now the actual input conductance. Figures 15.4 and 15.5 illustrate two methods of accomplishing the desired transformation. The former uses a step-down transformer, the latter a tapped coil. In addition, the latter taps the input on the coil, thus permitting the use of higher inductance, with consequent better Q.

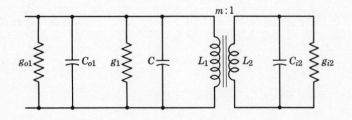

Fig. 15.4 Use of transformer to provide impedance matching.

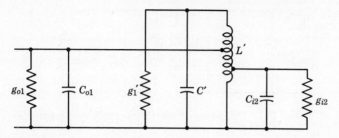

Fig. 15.5 Use of autotransformer to improve tuned circuit inductance and capacitance values.

15.4 Double-Tuned Coupling*

Figure 15.6 shows the equivalent circuit of a source and load coupled by a double-tuned circuit, having primary and secondary inductances L_1 and L_2 respectively, and associated tuning capacitances and leakage conductances. The source and load capacitances and conductances are as before, except that the secondary values are not transformed.

As before we can combine elements, resulting in the simplified circuit of Fig. 15.7. The conductances g_1 and g_{o1} have been combined into one conductance, taken as equal to xg_1; similarly, g_2 and g_{i2} are combined into yg_2. Thus, x and y represent factors by which the unloaded Q of the primary and secondary have been decreased by their respective loadings.

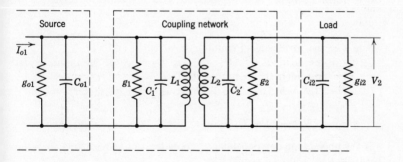

Fig. 15.6 Double-tuned circuit with output and input loadings.

* Ref. 4, end of chapter.

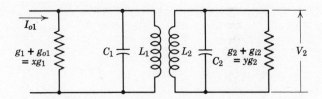

Fig. 15.7 Composite double-tuned circuit with loads.

As before, the power available is

$$P_1 = \frac{|I_{o1}|^2}{4g_{o1}} \qquad (15.13)$$

The power delivered to the load is

$$P_2 = |V_2|^2 g_{i2} \qquad (15.22)$$

The power transfer efficiency is, therefore,

$$\eta = 4g_{o1}g_{i2}\left|\frac{V_2}{I_{o1}}\right|^2 \qquad (15.23)$$

The factor V_2/I_{o1} is the magnitude of the open-circuit forward transfer impedance z_{21}. This transfer impedance has the following value:

$$|z_{21}| = \frac{\mu}{(1+\mu^2)\sqrt{xg_1yg_2}\sqrt{1 - \theta\rho^2/(1+\mu^2) + \rho^4/(1+\mu^2)^2}} \qquad (15.24)$$

In eq. 15.24 a number of new parameters have been introduced:

$$\mu = k\sqrt{Q_1Q_2}$$

where k = the coefficient of coupling between L_1 and L_2,
Q_1, Q_2 = the loaded Qs ($Q_1 = Q_0/x$, $Q_2 = Q_0/y$, assuming equal unloaded Qs for both coils, Q_0),
$\theta = 2(\mu^2 - b/2)/(1+\mu^2)$ (the so-called shape factor),
$b = Q_1/Q_2 + Q_2/Q_1$
$\rho = 2(f/f_0 - 1)\sqrt{Q_1Q_2}$.

ρ is the only frequency-dependent parameter and at resonance ($f = f_0$) becomes zero, and $|z_{21}|$ reduces to:

$$|z_{21}|_0 = \frac{\mu}{(1+\mu^2)\sqrt{xg_1yg_2}} \qquad (15.25)$$

The selectivity curve may be obtained from the ratio of z_{21} at the desired off-resonant frequency to its value at resonance:

$$\left|\frac{V}{V_0}\right| = \left[\sqrt{1 - \frac{\theta\rho^2}{1+\mu^2} + \frac{\rho^4}{(1+\mu^2)^2}}\right]^{-1} \quad (15.26)$$

The shape factor θ determines the shape of the selectivity curve. At $\theta = 0$, we have so-called transitional coupling, or maximally flat response. Above this value, a double-humped response will occur, whereas below it, the curve will narrow to that determined by the individual circuit Q.

The transfer efficiency can be expressed in terms of the above parameters and the loading factors x and y:

$$\eta = \frac{4(2-\theta)(x^2+y^2+\theta xy)(x-1)(y-1)}{(x+y)^4} \quad (15.27)$$

This will be maximized for the condition $x = y$, implying equal loaded Qs. The value of this maximum efficiency is:

$$\eta_{\max} = \left(1 - \frac{\theta^2}{4}\right)\left(1 - \frac{1}{x}\right)^2 \quad (15.28)$$

The value of x to produce this condition is:

$$x = \frac{Q_0 B_n' \sqrt{2-\theta}}{2\sqrt[4]{n}} \sqrt{\left(1+\frac{\theta^2}{4n}\right)^{\frac{1}{2}} - \frac{\theta}{2\sqrt{n}}}$$

The coefficient of coupling is:

$$k = \frac{x}{Q_0}\sqrt{\frac{2+\theta}{2-\theta}} \quad (15.30)$$

The inductance values are:

$$L_1 = \frac{x-1}{2\pi f_0 Q_0 g_{o1}} \quad (15.31)$$

$$L_2 = \frac{x-1}{2\pi f_0 Q_0 g_{i2}} \quad (15.32)$$

15.5 Transitional Coupling

At transitional coupling $\theta = 0$, and the above equations reduce to:

$$x_T = \frac{Q_0 B_n'}{\sqrt{2}\sqrt[4]{n}} \tag{15.33}$$

$$k_T = \frac{B_n'}{\sqrt{2}\sqrt[4]{n}} \tag{15.34}$$

$$\eta_{\max} = \left(1 - \frac{\sqrt{2}\sqrt[4]{n}}{Q_0 B_n'}\right)^2 \tag{15.35}$$

$$L_1 = \frac{(Q_0 B_n'/\sqrt{2}\sqrt[4]{n}) - 1}{2\pi f_0 Q_0 g_{o1}} \tag{15.36}$$

$$L_2 = \frac{(Q_0 B_n'/\sqrt{2}\sqrt[4]{n}) - 1}{2\pi f_0 Q_0 g_{i2}} \tag{15.37}$$

15.6 Procedure for Designing Double-Tuned Circuits

The following design procedure and design curves will facilitate the application of the foregoing analysis to actual design of double-tuned coupling circuits. This procedure and the curves are based on the assumption that the condition for optimizing transfer efficiency is attained, i.e., that $x = y$, and the two loaded Qs are equal. It is also assumed that the bandwidth and shape are specified, along with the center frequency, that the terminating loadings are known, and that we can use some realizable figure for the unloaded Q of the coils.

As an example in illustrating the application of this procedure, we will design a double-tuned circuit to have the following properties:

$$f_0 = 455 \text{ kc} \qquad \text{6-db bandwidth} = 5 \text{ kc} \qquad Q_0 = 160$$

$$\text{Output resistance} = 15{,}000 \text{ ohms} \qquad g_{o1} = 6.67 \times 10^{-5}$$

$$\text{Input resistance} = 1200 \text{ ohms} \qquad g_{i2} = 8.33 \times 10^{-4}$$

1. Calculate the normalized bandwidth:

$$B_n = \tfrac{5}{455} = 1.1 \times 10^{-2}$$

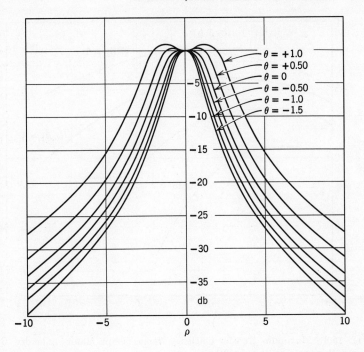

Fig. 15.8 Double-tuned coupling; selectivity curves versus shape factor θ.

2. Select desired shape of curve. Considerations: desired flatness at peak, skirt selectivity, maximum transfer efficiency. Figures 15.8–15.10 are used in making this decision. Figure 15.8 shows the shape of the selectivity curves as a function of the shape factor θ. As mentioned before, $\theta = 0$ produces maximum flatness, a greater value producing double-humped response, lower values producing sharper peaks and closer skirts. Figures 15.9 and 15.10 indicate how the maximum transfer efficiency varies with shape factor, for specified bandwidth at 3 db and 6 db respectively. Referring, for example, to Fig. 15.10, it will be seen that, for $Q_0 B_1 = 2$ or 3, the efficiency peaks at around $\theta = -1.0$, indicating that maximum gain is actually obtained with the coupling considerably below transitional, and if a flatter response is required it must be accompanied by loss of gain.

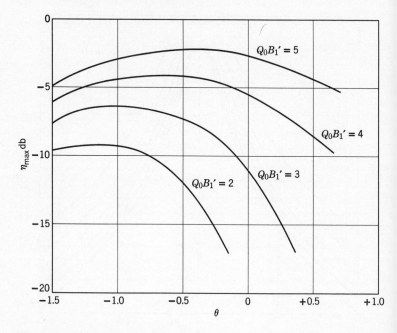

Fig. 15.9 Maximum transfer efficiency versus shape factor, unloaded Q and 3-db normalized bandwidth.

In our example, let us assume that we want to have maximum gain and choose a value of $\theta = -1.0$.

3. Determine the value of ρ. Referring to Fig. 15.8, for $\theta = -1.0$, we obtain, at the 6-db down points a value of ρ of 1.3.
4. Having chosen θ, and knowing the specified bandwidth number n (3 in this example, for 6 db), find $x/Q_0 B_n'$ from Fig. 15.11. In this example this turns out to be 0.792.
5. Calculate x: $Q_0 B_n' = 160 \times 5 \div 455 = 1.758$.

$$x = 0.792 \times 1.758 = 1.39$$

6. Obtain μ for chosen θ from Fig. 15.12. For $\theta = -1.0$, $\mu = 0.577$.
7. Calculate k: $k = x\mu/Q_0 = 1.39 \times 0.577 \div 160 = 0.005$ or 0.5%.
8. Obtain $\eta_{\max}$ from Fig. 15.13: for $x = 1.39$, $\theta = -1.0$, $\eta_{\max} = 0.060$.

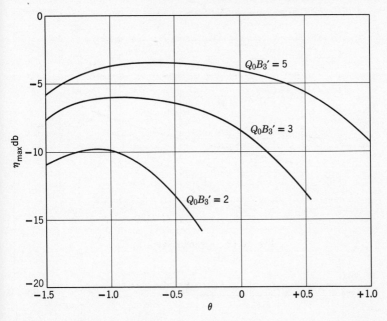

Fig. 15.10 Maximum transfer efficiency versus shape factor, unloaded Q and 6-db normalized bandwidth.

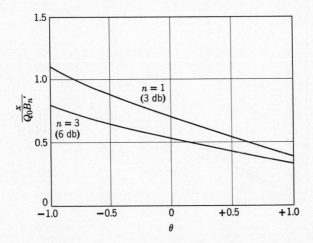

Fig. 15.11 Loading as a function of normalized bandwidth and shape factor.

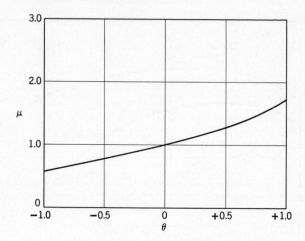

Fig. 15.12 μ versus shape factor.

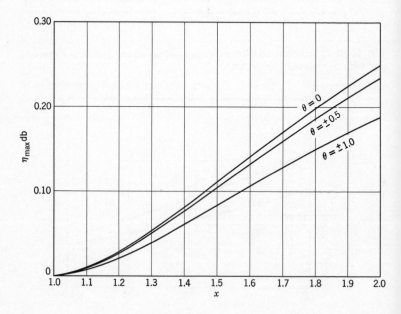

Fig. 15.13 Maximum transfer efficiency versus loading and shape factor.

9. Calculate inductances:

$$L_1 = \frac{x-1}{2\pi f_0 Q_0 g_{o1}} = 12.8 \times 10^{-6}$$

$$L_2 = \frac{x-1}{2\pi f_0 Q_0 g_{i2}} = 1.02 \times 10^{-6}$$

10. Calculate capacitors:

$$C_1 = \frac{1}{\omega^2 L_1} = 9600 \times 10^{-12} \qquad C_2 = \frac{1}{\omega^2 L_2} = 120{,}000 \times 10^{-12}$$

The above values of capacitors are unrealistically large; in actual practice we would use larger inductances and smaller capacitances, the limitation being primarily one of such factors as range of adjustment of inductors or capacitors, stray capacitances, etc.

Figure 15.14 shows the final circuit, designed in accordance with the above procedure. In this circuit, identical coils having inductances of 1.28 mh are used, with the equivalent of the calculated lower inductances being obtained by taps. The 0.1 tap on the primary indicates that the coil is tapped at one-tenth of its total turns, producing the equivalent of 12.8 μh for the input load. Similarly, the secondary load is tapped at about $\frac{1}{35}$th of its turns to produce 1.02 μh. This permits reducing the capacitors to 96 pf. (In this example, the input and output capacitances were assumed to be zero, otherwise their values should be subtracted from the calculated total before transforming.) The shunt

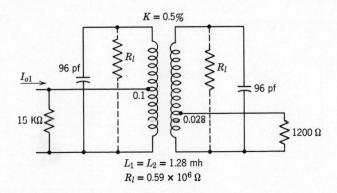

Fig. 15.14 Example of double-tuned circuit design.

resistors across the coils are shown dotted to indicate that these are the values required for leakage resistances to produce the stipulated unloaded Qs of 160, and are not externally added resistors.

REFERENCES

1. Linvill, J. G. and J. F. Gibbons, *Transistors and Active Circuits*, McGraw-Hill, New York, 1961, pp. 407–428.
2. Hunter, L. P., *Handbook of Semiconductor Electronics*, McGraw-Hill, New York, 1962, pp. 12-40–12-61.
3. Hurley, R. B., *Junction Transistor Electronics*, Wiley, New York, 1958, pp. 271–293.
4. Shea, R. F. (Ed.), *Transistor Circuit Engineering*, Wiley, New York, 1957, pp. 160–196.

16

Wide-Band Amplifiers

16.1 Introduction

In the previous chapter the design of narrow-band tuned amplifiers was described. In this chapter, the extension of high-frequency design principles to amplifiers having bandwidths comparable to the center frequency will be discussed.

Three major techniques are utilized, with variations, in the design of wide-band amplifiers: high-frequency peaking by means of inductance; over-all negative feedback; and distributed amplifiers. These three techniques will be discussed in some detail in the following sections.

16.2 High-Frequency Peaking

This technique is widely used in electron-tube video amplifiers and is also applicable with transistors. The basic principle involves increasing the effective load seen by the first stage at higher frequencies by the use of inductances in the coupling network. Figure 16.1 illustrates the technique, showing the a-c impedances involved in high-frequency amplification. The impedances of the batteries are assumed to be zero, also the impedances of coupling capacitors between stages and of any bias networks in the emitter leads. As will be seen later, this does not apply at low frequencies. Resistor R_1 is the low-frequency collector load resistor, in series with which an inductance L_1 has been connected to raise effectively the load impedance at high frequencies. A second inductance L_2 is interposed between the two transistors, which partially

156   TRANSISTOR APPLICATIONS

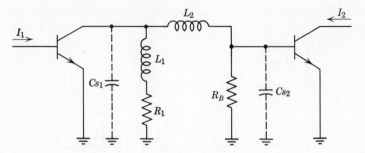

Fig. 16.1   Use of high-frequency peaking coils.

tunes the input and output capacitances. C_{s1} and C_{s2} represent stray capacitances. R_B is the equivalent resistance of the bias network on the base of the second transistor.

Figure 16.2 shows the equivalent circuit for this arrangement. The first transistor is represented by the current generator $-h_{fe}I_{b1}$ (the justifiable assumption is made that the load impedance is low enough so that the transistor can be assumed to be working into essentially a short-circuit) and by the output conductance g_o and capacitance C_o (into which the stray capacitance C_{s1} has been absorbed). Similarly, the second transistor is represented by the input conductance g_i and capacitance C_i (which also includes C_{s2}). V_{i2} is the output voltage appearing across the network output terminals and, hence, at the input to the second transistor.

The network of Fig. 16.2 can be analyzed by any of the network

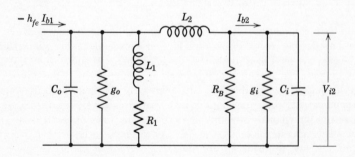

Fig. 16.2   Equivalent circuit at high frequencies.

methods described earlier. One of the most useful forms is the z representation, with the network represented by its transfer impedance z_{21}, given by the equation

$$z_{21} = -\frac{V_{i2}}{h_{fe}I_{b1}} \tag{16.1}$$

The over-all current amplification of the combination of two transistors and coupling network is obtained from the following relationship:

$$A_i = \frac{I_{c2}}{I_{b1}} = \frac{-(h_{fe})^2 z_{21}}{Z_{ie}} \tag{16.2}$$

In the above equation it is assumed that the h_{fe} of the two transistors are alike and that, furthermore, the output load on the second transistor is also low enough so that it can be assumed working into virtually a short-circuit. The negative sign is the result of our current direction convention. Z_{ie} is the input impedance of the second transistor.

In obtaining the values for the output admittances and input impedances of the transistors, the equations of Chapter 14 are used. The pertinent ones are:

$$Z_{ie} \simeq r_b' + \frac{r_\epsilon}{1/[(h_{fe})_0 + 1] + j\omega/\omega_{ab}} \tag{16.3}$$

$$\frac{1}{Z_{oe}} \simeq \frac{(g_c + j\omega C_c)(1 + j\omega/\omega_{ab})}{1/[(h_{fe})_0 + 1] + j\omega/\omega_{ab}} \tag{16.4}$$

$$h_{fe} \simeq \frac{(h_{fe})_0}{1 + j[(h_{fe})_0 + 1]\omega/\omega_{ab}} \tag{16.5}$$

As an illustrative example, take the 2N1613 transistor, operating at the rated emitter current and collector voltage, and use the following values for the parameters:

$r_\epsilon = 25$ ohms $r_b' = 100$ ohms $(h_{fe})_0 = 55$ $f_{ab} = 130 \times 10^6$ cps
$C_c = 18 \times 10^{-12}$ µfd $g_c = 0.16 \times 10^{-6}$ mho

158   TRANSISTOR APPLICATIONS

The complex input impedance, as calculated from eq. 16.3 and its reciprocal, the input admittance, are:

f	Z_{ie}	Y_{ie}
0	1,500	0.667×10^{-3}
0.1 Mc	$1,470 - j59$	$(0.680 + j0.027)10^{-3}$
1	$1,280 - j505$	$(0.674 + j0.265)10^{-3}$
2	$903 - j689$	$(0.667 + j0.510)10^{-3}$
4	$452 - j605$	$(0.790 + j1.060)10^{-3}$
10	$172 - j308$	$(1.384 + j2.48)10^{-3}$

From the above it is seen that g_i remains essentially constant up to about 2 Mc and rises above that value. The value of C_i, as calculated from the reactive portion of Y_{ie}, remains relatively constant at an average value of 41.5 pf. An assumed stray capacitance of 3.5 pf is added to this in the following calculations, making a total for C_i of 45 pf.

In similar manner the output admittance Y_{oe} is obtained from eq. 16.4. A stray capacitance of 3.5 pf was also added to C_o and the total values of $g_o + j\omega C_o$ are given below.

f	$Y_{oe} = g_o + j\omega C_o$
0	8.95×10^{-6}
0.1 Mc	$(36.9 + j648)10^{-6}$
1	$(2,310 + j5502)10^{-6}$
2	$(4,160 + j7594)10^{-9}$
4	$(7,300 + j6968)10^{-6}$
10	$(14,050 + j4620)10^{-6}$

Finally, the variation of h_{fe} with frequency is:

f	h_{fe}
0	55
0.1 Mc	$54.9 - j2.4$
1	$46.2 - j19.9$
2	$31.1 - j25.8$
4	$14.7 - j24.4$
10	$2.8 - j12.1$

The equation for the transfer impedance z_{21} has been calculated for the circuit of Fig. 16.2 and is:

$$z_{21} = \left[\frac{1}{R_B} + g_i + j\omega C_i + \left(g_o + j\omega C_o + \frac{1}{R_1 + j\omega L_1} \right) \left(1 - \omega^2 L_2 C_i + j\omega g_i L_2 + \frac{j\omega L_2}{R_B} \right) \right]^{-1} \quad (16.6)$$

The roles of the various components of the circuit are not easily visualized from the above equation. It is apparent that the bias resistor R_B produces loss. Also, increasing L_1 will increase gain, since it reduces the loading effect of resistor R_1, although there will be a practical limit to this improvement, imposed by the distributed capacitance of the choke. The effect of L_2 is more obscure, since there is obviously a resonance condition possible between the two inductances and the two capacitances. The condition for zero phase shift occurs when the reactive components of eq. 16.6 cancel. This produces the following

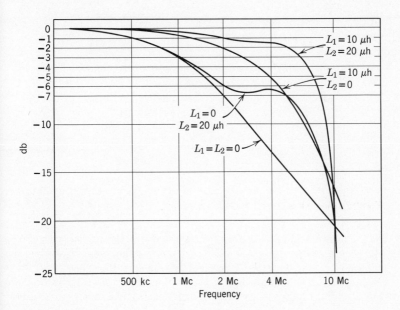

Fig. 16.3 High-frequency response for various values of peaking coil inductances.

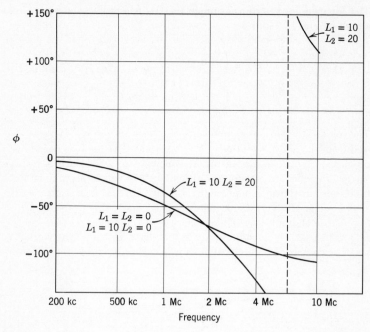

Fig. 16.4 Phase shift versus frequency for various values of coil inductances.

relationship between the inductances for this condition:

$$L_2 = \frac{\omega^2 L_1{}^2(C_o + C_i) - L_1 + R_1{}^2(C_o + C_i)}{(\omega^2 C_i C_o - g_i g_o - g_o/R_B)(\omega^2 L_1{}^2 + R_1{}^2) - \omega^2 L_1 C_i - g_i R_1 - R_1/R_B} \quad (16.7)$$

The curves of Fig. 16.3 have been plotted to illustrate the effects of the peaking inductances. A value of R_B of 20,000 ohms was assumed. Without peaking coils, the 3-db point is 1 Mc and the fall-off is at the customary rate of 6 db per octave. Addition of a 10 μh coil for L_1 alone raises the 3-db point to nearly 3 Mc, at the expense of a somewhat greater fall-off rate. Using L_2 alone produces the apparent resonance effect at around 4 Mc. The combination of $L_1 = 10$ μh and $L_2 = 20$ μh produces a bandwidth of about 6 Mc, with a very rapid falloff. The peak at 4 Mc could be raised to the 0-db level, if desired, with a small dip at about 3 Mc.

Figure 16.4 shows the phase shift for the conditions where no peaking coils are used—or only the 10-μh coil—and the condition for maximum bandwidth, using both coils. It is apparent that the latter combination produces considerable phase shift, which, furthermore, is not linear with frequency, thus this combination would not be satisfactory for a pulse amplifier. Such amplifiers rely usually on negative feedback to obtain proper pulse response, and this type of wide-band amplifier will be discussed later in this chapter.

16.3 Low-Frequency Compensation

The same technique is employed with transistors as with tubes, i.e., increasing the effective load impedance at low frequencies by using a frequency-selective R-C combination in series with the high-frequency

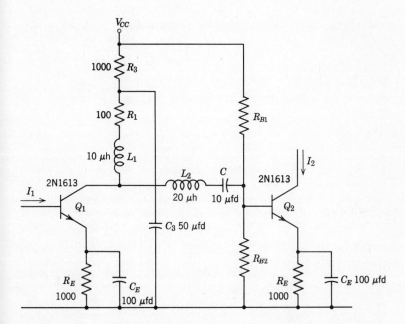

Fig. 16.5 Complete circuit showing peaking coils and low-frequency emitter impedances and compensating load circuit.

load resistance R_1. This technique is illustrated in the circuit of Fig. 16.5, which shows the complete circuit of a two-stage amplifier intended to have a bandwidth of about 6 Mc. It incorporates the high-frequency peaking coils described above and, in addition, has several network elements which were negligible at the high-frequency end. The coupling capacitor C now has considerable reactance; furthermore, the bias networks in the emitter leads will produce equivalent impedances in the input circuits, multiplied by $h_{fe} + 1$. The net result of these impedances is to reduce radically the low-frequency response (see Figs. 9.5 and 9.6, Chapter 9). To compensate for this effect, the resistor R_3 is inserted in series with R_1, by-passed at high frequencies by the capacitor C_3. At low frequencies, where the reactance of C_3 becomes comparable to the resistance R_3, the net load impedance increases and the gain is restored. As a rough rule of thumb, the compensating network should have approximately the same time constant as the emitter bias network.

The transfer impedance for the low-frequency case may be obtained from the following equation:

$$z_{21} = \left[\frac{1}{R_B} + \frac{1}{Z_{ie}} + \left(g_o + \frac{1 + j\omega R_3 C_3}{R_1 + R_3 + j\omega R_1 R_3 C_3} \right) \left(1 + \frac{1}{j\omega C Z_{ie}} + \frac{1}{j\omega C R_B} \right) \right]^{-1} \quad (16.8)$$

[Z_{ie} includes $Z_E(h_{fe} + 1)$, see Section 8.6, Chapter 8]. Using the above equation for z_{21}, the over-all amplification may again be calculated, as before:

$$A_i = (h_{fe})^2 \frac{z_{21}}{Z_{ie}} \quad (16.9)$$

Figure 16.6 shows the over-all performance of this circuit, calculated from the above equations. The 3-db points are approximately 50 cycles and 6 Mc. The low-frequency response can be further raised by increasing the values of the various capacitors. Another expedient is to use breakdown diodes as coupling devices and/or as biasing means, thus avoiding the high impedances produced by the use of by-pass and coupling capacitors.

As an example of the current state of the art of high-frequency wide-band amplifier design, principles similar to those described above were applied to the design of an L-band amplifier using experimental transistors with cutoff frequencies in the gigacycle range. The resulting design had a gain of about 32 db, and a response essentially flat from about 1.1 to 1.6 gigacycles.*

* Reference 3, end of chapter.

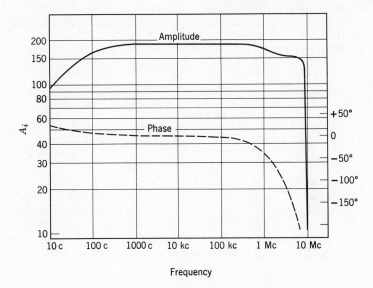

Fig. 16.6 Curves of over-all gain and phase shift versus frequency.

16.4 Use of Feedback in Broad-Band Amplifiers

As with electron tubes, gain can be exchanged for bandwidth in transistor amplifiers, and feedback arrangements are frequently used for this purpose. In general this feedback takes one of two forms, with occasionally both being employed. Figures 16.7 and 16.8 illustrate these. In Fig. 16.7 feedback occurs from the emitter of the second transistor to the base of the first, and the over-all amplification is determined by the resistors R_1 and R_2. Normally R_1 is very low, on the order of 10-25 ohms, R_2 high, on the order of 1000-2000 ohms. In a typical circuit employing this method of feedback, and using 2N502 transistors, a bandwidth of 18 Mc was obtained, with gain-bandwidth product of about 1300.

In Fig. 16.8 the feedback is from the output collector to the first emitter, and the amplification is determined by the resistors R_1 and R_2 again. In both these circuits the feedback reduces the dependence of gain on transistor parameters and produces an amplifier relatively independent of environmental conditions, as well as of transistor variations.

Figure 16.9 shows a circuit incorporating both methods of feedback.[*]

[*] Reference 10, end of chapter.

164 TRANSISTOR APPLICATIONS

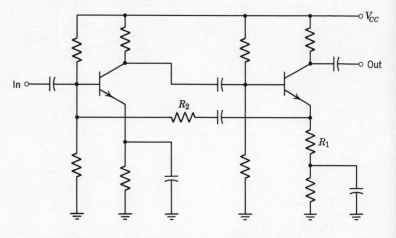

Fig. 16.7 Use of feedback from second emitter to first base.

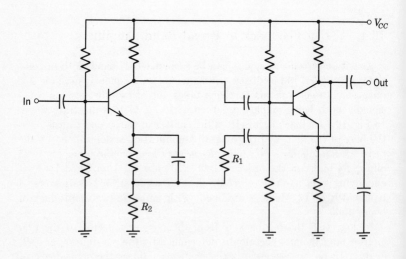

Fig. 16.8 Use of feedback from output to first emitter.

WIDE-BAND AMPLIFIERS 165

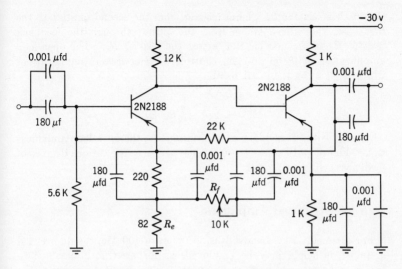

Fig. 16.9 Circuit employing both methods for feedback.

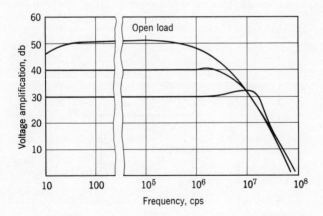

Fig. 16.10. Frequency response of circuit of Fig. 16.9.

One path is via the 22-kilohm resistor from the second emitter to the first base, the other path is from the output through the feedback resistor R_f to the first emitter, across the resistor R_e. The open-loop amplification is 48–50 db. R_f controls the closed-loop amplification, which is given by the relationship:

$$A_{vc} = \frac{A_{vo}}{1 - A_{vo}R_e/R_f} \qquad (16.10)$$

where A_{vc} is the closed-loop amplification, A_{vo} the open-loop amplification. Figure 16.10 shows the frequency response for various degrees of feedback.

16.5 Distributed Amplifiers

For extremely wide bandwidths, in excess of 100 Mc, the distributed amplifier principle may be employed, as for electron tubes.* Two transmission lines are used, one connected to the input, the other to the output, and the transistors are connected between the two lines. One section of such an amplifier is shown in Fig. 16.11. This section comprises a high-frequency transistor, such as the 2N706 or 2N753, connected between the two transmission lines (half of each section of line as shown). A network, comprised of a resistor and capacitor in parallel, is

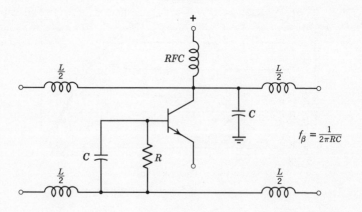

Fig. 16.11 Section of distributed amplifier. (From Ref. 2.)

* See Ref. 2, end of chapter.

WIDE-BAND AMPLIFIERS   167

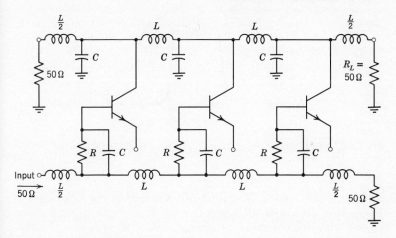

Fig. 16.12 Single-stage distributed amplifier, incorporating three sections.

inserted in the base lead. This produces a rise in base current with frequency, approximately compensating for the fall-off of h_{fe}.

A single stage of a distributed amplifier incorporates a number, e.g., three or four, of sections in one assembly. Such a three-section stage is shown in Fig. 16.12. This amplifier is designed for 50-ohm line impedance, hence has an input and output impedance of this value. Thus, a number of such stages may be cascaded to increase the over-all gain. Such a four-stage amplifier is described in Ref. 2. The frequency response is shown in Fig. 16.13. The response of this amplifier to pulses

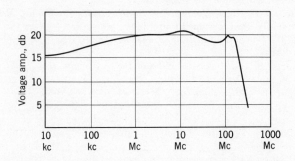

Fig. 16.13 Frequency response of distributed amplifier.

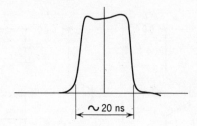

Fig. 16.14   Output pulse waveform from distributed amplifier.

is shown in Fig. 16.14. The combined rise time of the four stages is about 3.5 nanoseconds when adjusted for no overshoot, or 2.5 nanoseconds with 10% overshoot.   Output capability is ± 1 v or 2 v peak-to-peak. The gain is about 20 db up to 200 Mc.

REFERENCES

1. Bénéteau, Paul J., and Larry Blaser, The Design of Distributed Amplifiers Using Silicon *Mesa* Transistors, *Technical Paper TP-13*, Fairchild Semiconductor Corp.
2. Bénéteau, Paul J., and Larry Blaser, A 175 MC Distributed Amplifier Using Silicon *Mesa* Transistors, *Application Data APP-14*, Fairchild Semiconductor Corp.
3. Hamasaki, J., *A Wideband High-Gain Transistor Amplifier at L-Band*, Paper No. THAM 5.1, presented at the 1963 International Solid-State Circuits Conference, Philadelphia, Pa. (Feb. 21, 1963). Abstract on pp. 46–47 of Digest of technical papers.
4. Hunter, L. P. (Ed.), *Handbook of Semiconductor Electronics*, McGraw-Hill, New York, 1962, pp. 12-30–12-60.
5. Kolk, P., *Design of Wideband Transistor Amplifiers*, RCA Application Note SMA-7.
6. Schwartz, S. (Ed.), *Selected Semiconductor Circuits Handbook*, Wiley, New York, 1960, pp. 4-1 – 4-49.
7. Shea, R. F. (Ed.), *Transistor Circuit Engineering*, Wiley, New York, 1957, pp. 199–219.
8. Simmons, C. D. and P. G. Thomas, *The Design of High-Frequency Diffused-Base Transistor Amplifiers*, Sprague Electric Co., Application Note No. 38,009.
9. RCA Silicon VHF Transistors, Application Guide 1CE-228.
10. Texas Instruments, Inc., *Transistor Circuit Design*, McGraw-Hill, New York, 1963, pp. 253–271

17

Sinusoidal Oscillators

17.1 Introduction

The oscillators to be described in this chapter are called sinusoidal to distinguish them from the various types of relaxation oscillators to be covered in the next chapter. The approach here is to apply those principles which have already been developed for the design of amplifiers, treating oscillators as amplifiers terminated in their own input impedances and which produce unity amplification without phase shift under these conditions. Several typical forms of oscillator are discussed and these principles are applied to their solution.

17.2 The Oscillator as a Closed-Loop Amplifier

Consider the elementary block-diagram of Fig. 17.1. A is a voltage amplifier, N a network having certain properties. When loaded with an impedance Z_i, the network N will have an input impedance Z_l. Furthermore, the voltage transfer ratio of the network is V_i/V_n. We define two parameters of the network:

$$A_N = \text{the voltage ratio} = \frac{V_i}{V_n} \text{ as shown} \qquad (17.1)$$

$$M_N = \text{the impedance ratio} = \frac{Z_l}{Z_i} \text{ as shown} \qquad (17.2)$$

'Another way of expressing this would be that, when terminated in an

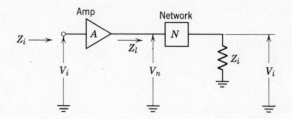

Fig. 17.1 Block diagram of oscillator, showing amplifier and coupling network.

impedance Z_i, the network would have an input impedance $M_N Z_i$ and a voltage ratio A_N.

The amplifier will, in turn, have a voltage amplification given by

$$A_v = \frac{-h_{21} Z_l}{h_{11} + \Delta^h Z_l} \tag{17.3}$$

If $\Delta^h Z_l \ll 1$, eq. 17.3 can be approximated by

$$A_v \simeq \frac{-h_{21} Z_l}{h_{11}} \tag{17.4}$$

The input impedance of the amplifier is:

$$Z_i = \frac{h_{11} + \Delta^h Z_l}{1 + h_{22} Z_l} \tag{17.5}$$

If Z_l is low, as is usually the case in oscillators, this approximates to

$$Z_i = h_{11} \tag{17.6}$$

Combining eqs. 17.1, 17.2, 17.4, and 17.6, and stipulating that the overall amplification $A_v A_N$ must equal $1 + j0$ for oscillations to occur, we arrive at the expression:

$$A_N M_N = \frac{-1}{h_{21}} \tag{17.7}$$

This simple expression permits analysis of any oscillator circuit when it is broken into its component amplifier and coupling network. The following section will illustrate the application of this technique.

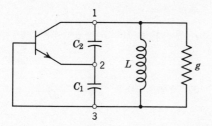

Fig. 17.2 Basic circuit of Colpitts oscillator.

17.3 The Colpitts Oscillator

Figure 17.2 shows the basic circuit of the transistor version of the well-known Colpitts oscillator, with d-c bias elements omitted for simplicity. The conductance g represents the various losses, not only in the tuned circuit but also for power delivered to a load.

Figure 17.3 shows the coupling network elements arranged to facilitate analysis as a network. The network is terminated in the transistor impedance Z_i, which is assumed to be given accurately enough by the short-circuit impedance h_{ie}.

The easiest method of analyzing this network is by means of y parameters. These can be obtained as follows: y_{11} is input admittance, output short-circuited. By inspection this is seen to be the net admittance of C_2, L and g in parallel, or

$$y_{11} = g + \frac{1 - \omega^2 L C_2}{j \omega L} \tag{17.8}$$

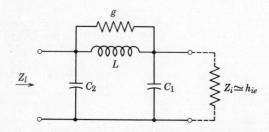

Fig. 17.3 Coupling network of Colpitts oscillator.

172   TRANSISTOR APPLICATIONS

Similarly,
$$y_{22} = g + \frac{1 - \omega^2 L C_1}{j\omega L} \tag{17.9}$$

and
$$y_{12} = y_{21} = -\left(g + \frac{1}{j\omega L}\right) \tag{17.10}$$

The determinant,
$$\Delta^y = \frac{C_1 + C_2}{L} - \omega^2 C_1 C_2 + j\omega g(C_1 + C_2) \tag{17.11}$$

We can now apply network equations to determine the input impedance and voltage ratio of this network when terminated in an impedance h_{ie}. Input impedance of network, Z_l,

$$Z_l = \frac{1 + y_{22} h_{ie}}{y_{11} + \Delta^y h_{ig}}$$

$$= \frac{1 + [g + (1 - \omega^2 L C_1)/j\omega L] h_{ie}}{g + (1 - \omega^2 L C_2)/j\omega L + [(C_1 + C_2)/L - \omega^2 C_1 C_2 + j\omega g(C_1 + C_2)] h_{ie}} \tag{17.12}$$

$$M_N = \frac{Z_l}{h_{ie}}$$

Voltage ratio:
$$A_N = \frac{-y_{21} h_{ie}}{1 + y_{22} h_{ie}}$$

$$= \frac{(g + 1/j\omega L) h_{ie}}{1 + [g + (1 - \omega^2 L C_1)/j\omega L] h_{ie}} \tag{17.13}$$

Multiplying A_N and M_N:

$$A_N M_N = \frac{g + 1/j\omega L}{g + (1 - \omega^2 L C_2)/j\omega L + [(C_1 + C_2)/L - \omega^2 C_1 C_2 + j\omega g(C_1 + C_2)] h_{ie}} \tag{17.14}$$

This is the quantity which must be equated to $-1/h_{fe}$. Thus,

$$-h_{fe}\left(g + \frac{1}{j\omega L}\right) = g + \frac{1 - \omega^2 L C_2}{j\omega L} + \left[\frac{C_1 + C_2}{L} - \omega^2 C_1 C_2 + j\omega g(C_1 + C_2)\right] h_{ie} \tag{17.15}$$

Substituting the complex forms

$$h_{fe} = h_{feR} + jh_{feI} \quad \text{and} \quad h_{ie} = h_{ieR} + jh_{ieI}$$

where the added subscripts R and I indicate real and imaginary components, and separating the reals of the complete equation from the imaginary terms, we finally obtain two simultaneous equations, both of which must be satisfied for oscillation to occur:

$$g(1 + h_{feR}) + \frac{h_{feI}}{\omega L} + h_{ieR}\left(\frac{C_1 + C_2}{L} - \omega^2 C_1 C_2\right)$$
$$- \omega g h_{ieI}(C_1 + C_2) = 0 \quad (17.16)$$

$$\frac{h_{feR}}{\omega L} - g h_{feI} + \frac{1}{\omega L} - \omega C_2 - \omega g h_{ieR}(C_1 + C_2)$$
$$- h_{ieI}\left(\frac{C_1 + C_2}{L} - \omega^2 C_1 C_2\right) = 0 \quad (17.17)$$

An approximate solution may be obtained by neglecting g, for which condition eqs. 17.16 and 17.17 reduce to the following:

$$h_{feR} = \omega h_{ieI}(C_1 + C_2 - \omega^2 L C_1 C_2) + \omega^2 L C_2 - 1 \quad (17.18)$$

$$h_{feI} = \omega h_{ieR}(\omega^2 L C_1 C_2 - C_1 - C_2) \quad (17.19)$$

Combining these equations and solving,

$$\omega^2 L C_2 = 1 + h_{feR} + \frac{h_{ieI} h_{feI}}{h_{ieR}} \quad (17.20)$$

and

$$C_1 = \frac{h_{ieR} C_2 + h_{feI}/\omega}{h_{ieR} h_{feR} + h_{ieI} h_{feI}} \quad (17.21)$$

As proof of the validity of the assumptions on which the above approach is based, the values for a 1-Mc oscillator were computed, for a transistor having the following complex parameters:

$$h_{ie} = 170 - j30.7$$
$$h_{fe} = 0.031 - j1.18$$
$$h_{oe} = (260 + j226)10^{-6}$$
$$\Delta^h{}_e = 0.051 + 0.030 \quad (h_{re} \text{ assumed negligible})$$

A value of $\omega L = 150$ ohms was chosen arbitrarily, thus $L = 23.9$ μh. Putting the proper values in eqs. 17.20 and 17.21 and solving, we obtain:

$$C_1 = 900 \times 10^{-12} \qquad C_2 = 1320 \times 10^{-12}$$

Substituting these values in the equation for Z_l (eq. 17.12 with $g = 0$):

$$Z_l = -122.0 + j16.1$$

This is the load into which the transistor works. Now, substituting this value of Z_l into the equation for transistor input impedance:

$$Z_{ie} = \frac{h_{ie} + \Delta^h{}_e Z_l}{1 + h_{oe} Z_l}$$

and solving:

$$Z_{ie} = 169.5 - j30.5$$

This is obviously essentially the same as the original value of h_{ie}.

17.4 The Hartley Oscillator

This is another long-time tube favorite which has its transistor counterpart. The circuit is shown in Fig. 17.4, while Fig. 17.5 shows the coupling network. In this case the loss conductance g is omitted, and may most easily be included by incorporating it within the net input impedance Z_{ie} by considering it as an additional shunt across the transistor.

Although the analysis of this circuit is complicated by the presence of the mutual inductance M, the same approach can be used as for the Colpitts oscillator, namely, by analyzing the coupling network to obtain the factors A_N and M_N, using the short-circuit y parameters of the network. This has been done, resulting in the following solution, assuming

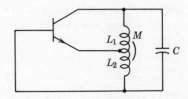

Fig. 17.4 Basic circuit of Hartley oscillator.

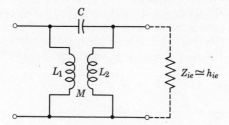

Fig. 17.5 Coupling network of Hartley oscillator.

again that the h_{re} and h_{oe} parameters have negligible effect, and that the input impedance can be assumed equal to h_{ie}:

$$j\omega M[(\omega^2 MC + \omega^2 L_2 C)(h_{fe} + 1) - h_{fe}]$$
$$- j\omega L_2[(\omega^2 MC + \omega^2 L_1 C)(h_{fe} + 1) - 1]$$
$$- h_{ie}(2\omega^2 MC + \omega^2 L_1 C + \omega^2 L_2 C - 1) = 0 \quad (17.22)$$

The complex forms must now be substituted for h_{fe} and h_{ie} and then the reals and imaginaries must be separated. This results in the following two simultaneous equations:

$$h_{feI}[\omega M(1 - \omega^2 MC) + \omega^3 L_1 L_2 C]$$
$$+ h_{ieR}(1 - \omega^2 L_1 C - \omega^2 L_2 C - 2\omega^2 MC) = 0 \quad (17.23)$$
$$\omega M h_{feR}(\omega^2 MC - 1) - \omega^2 L_1 L_2 C(h_{feR} + 1) + \omega^2 MC + \omega L_2$$
$$+ h_{ieI}(1 - \omega^2 L_1 C - \omega^2 L_2 C - \omega^2 MC) = 0 \quad (17.24)$$

Again, both equations must be satisfied for oscillation to occur.

17.5 The R-C Phase-Shift Oscillator

This form of oscillator is normally used only at fairly low frequencies, where the transistor parameters are essentially real numbers. Figure 17.6 shows the circuit of such an oscillator, and Fig. 17.7 shows the coupling network in similar manner to the other oscillators. The bias networks are shown here to emphasize that they can have considerable effect on the over-all circuit design, particularly the emitter bias network, since any phase shift it produces must be compensated for by the phase-shift network. If such an oscillator is to be used for wide-range coverage, with the frequency decided solely by the values of the capacitors C and resistors R, such bias network phase shift must be avoided.

The network of Fig. 17.7 can be analyzed to obtain the network parameters by the same technique as previously described for the other types of oscillators.

$$A_N = \frac{R_{ie}}{R - \dfrac{5}{\omega^2 R C^2} + \dfrac{6}{j\omega C} - \dfrac{1}{j\omega^3 R^2 C^3}} \qquad (17.25)$$

$$M_N = \frac{R - \dfrac{5}{\omega^2 R C^2} + \dfrac{6}{j\omega C} - \dfrac{1}{j\omega^3 R^3 C^3}}{R_{ie}\left[\dfrac{R}{R_l} - \dfrac{5}{\omega^2 R R_l C^2} - \dfrac{6}{j\omega R_l C} - \dfrac{1}{j\omega^3 R^2 R_l C^3} + 3 + \dfrac{4}{j\omega RC} - \dfrac{1}{\omega^2 R^2 C^2}\right]} \qquad (17.26)$$

Setting $A_N M_N = -1/h_{fe}$, we obtain:

$$h_{fe} = \frac{R}{R_l} - \frac{5}{\omega^2 R R_l C^2} + \frac{6}{j\omega R_l C} - \frac{1}{j\omega^3 R^2 R_l C^3} \\ + 3 + \frac{4}{j\omega RC} - \frac{1}{\omega^2 R^2 C^2} \qquad (17.27)$$

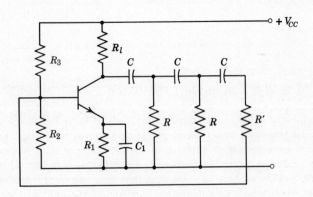

Fig. 17.6 Basic circuit of phase-shift oscillator.

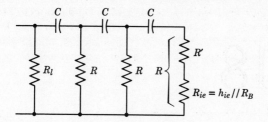

Fig. 17.7 Coupling network of phase-shift oscillator.

17.6 Other Types of Oscillators

The foregoing described a technique which is applicable to the analysis of many other types of oscillators, among which are such variations as the Clapp oscillator, where the inductance of the Colpitts oscillator is changed to a series-tuned circuit; tuned-base, tuned-emitter oscillator; combinations of positive and negative feedback, e.g., the arrangement where over-all positive feedback is combined with the use of a series-tuned circuit in the emitter. At frequencies away from the resonance of this tuned circuit, the emitter network introduces sufficient negative feedback to overcome the positive feedback, but at resonance the positive feedback overrides, thus oscillation will occur at the resonant frequency of the emitter circuit. Twin-T rejection filters can also be used in similar combinations of positive and negative feedback.

REFERENCES

1. Greiner, R. A., *Semiconductor Devices and Applications*, McGraw-Hill, New York, 1961, pp. 331–343.
2. Hunter, L. P. (Ed.), *Handbook of Semiconductor Electronics*, McGraw-Hill, New York, 1962, pp. 14-1–14-25.
3. Hurley, R. B., *Junction Transistor Electronics*, Wiley, New York, 1958, pp. 304–323.
4. Schwartz, S. (Ed.), *Selected Semiconductor Circuits Handbook*, Wiley, New York, 1960, pp. 5-1–5-40.
5. Shea, R. F. (Ed.), *Transistor Circuit Engineering*, Wiley, New York, 1957, pp. 221–241.
6. Texas Instruments, Inc., *Transistor Circuit Design*, McGraw-Hill, New York, 1963, pp. 307–320.

18

Relaxation Oscillators, Multivibrators

18.1 Introduction

In the previous chapter, oscillators were described whose mode of operation involved resonance of a tuned circuit. In this chapter we will discuss those oscillators where the phenomenon of negative resistance is used, in conjunction with some form of energy-storage element, to produce so-called relaxation oscillation. The various forms of multivibrators are typical of this class of transistor circuit, and find considerable application in switching, computing, logic, and similar applications, as well as for the generation of various nonsinusoidal wave shapes, e.g., square waves.

18.2 The Astable Multivibrator

This oscillator, also called a free-running multivibrator, is widely used for the generation of a variety of waveforms or as a frequency source where high purity of waveform is not required. It consists essentially of a symmetrical two-transistor circuit possessing a negative-resistance characteristic, biased so that its load line only intersects the unstable portion of the characteristic; hence the circuit has no stable operating point and will oscillate in a relaxation mode if coupled to an energy-storage device, such as a capacitor.

The analysis of such a circuit can be facilitated by opening the circuit at a chosen point and inserting a voltage source, then computing the resultant current as a function of this voltage. If the resultant curve

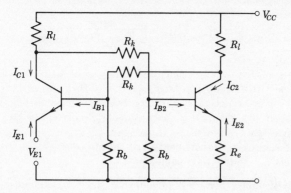

Fig. 18.1 Basic circuit of astable multivibrator, with voltage inserted in first emitter lead.

shows a region where the current decreases as the voltage increases, or vice versa (bearing in mind the directions of currents and voltages in transistor elements), we have a negative-resistance region and astable operation should be possible.

Figure 18.1 illustrates this technique. The circuit is symmetrical with the exception that the emitter resistor in the first transistor has been removed and a voltage has been inserted at this point instead. No capacitors are shown in this figure, since it is primarily intended to investigate the steady-state conditions. The capacitors are necessary in the final design to provide charge storage and determine the operating frequency.

The various currents and voltages can be calculated from a knowledge of the resistors and supply voltage. The emitter currents are given by the following equations (neglecting I_{CBO}):

$$I_{E1} = \frac{V_{CC}(1 - B) - V_{BE2}/k + BV_{BE1}/k + BV_{E1}/k(h_{FE} + 1)}{-h_{FE}R_l + B(R_k + R_l)} \tag{18.1}$$

$$I_{E2} = \frac{(h_{FE} + 1)(V_{BE2}/k - V_{CC}) - h_{FE}R_l I_{E1}}{[1 + R_e(h_{FE} + 1)/R_b](R_k + R_l) + R_e(h_{FE} + 1)} \tag{18.2}$$

where

$$k = \frac{R_b}{R_b + R_k + R_l}$$

and

$$B = \frac{[1 + (h_{FE} + 1)R_e/R_b](R_k + R_l) + R_e(h_{FE} + 1)}{h_{FE}R_l} \quad (18.3)$$

The base-emitter voltages V_{BE1} and V_{BE2} are those corresponding to their respective currents, I_{E1} and I_{E2}. Since they appear in the equations which determine these currents, estimates must first be made of their magnitudes, then, after the currents have been calculated, second-order corrections can be made, if necessary.

As an example of the application of the above equations, consider a design having the following values:

$R_e = 1.1$ kilohms $R_l = 3.4$ kilohms $R_b = 7.7$ kilohms
$R_k = 27.4$ kilohms $V_{CC} = 12$ v

First, let us assume a value of 20 for h_{FE}. Figure 18.2 gives the result of the calculation of I_{E1} for various values of V_{E1}, as the dotted curve. The two extremes of the curve represent the conditions where the first transistor is driven into either saturation or cutoff. Since the emitter current becomes *less* negative (or, in other words, more positive) as V_{E1}

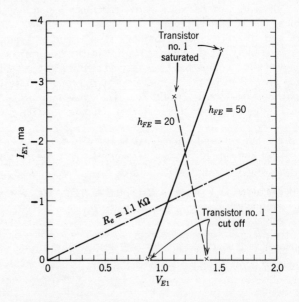

Fig. 18.2 Emitter current versus emitter voltage for circuit of Fig. 18.1, showing effect of different h_{FE}.

increases, this represents a positive resistance, and oscillation is therefore not possible with this low value of h_{FE}.

Let us now recalculate the performance with a value of $h_{FE} = 50$. The solid curve of Fig. 18.2 shows the results of this change. It will now be seen that the direction of the curve is reversed, and we now have a negative resistance region. Also shown on this figure is a line representing an emitter resistance of 1.1 kilohm. It will be noted that this resistance permits intersection well within the negative-resistance region; therefore, if such a resistor is now inserted in place of the voltage V_{E1}, we have a potentially active multivibrator. Since this value is also that inserted in the emitter of the second transistor, the multivibrator will be symmetrical. If we now add capacitors C_k across the resistors R_k and capacitors C_e across R_e, the circuit will oscillate with a frequency determined primarily by C_k and its associated resistors R_k, R_l, and R_b. The frequency will be between $1/2C_kR_b$ and $1/2C_kR_0$, where R_0 is the shunt combination of the three resistors, and will also depend upon such other factors as the degree of saturation, output voltage swing, etc.

Many techniques have been proposed for synthesizing astable multivibrators, as well as the other varieties, some with more rigor than others. For example, the factor I_{CBO} was neglected in the above analysis. In silicon transistors this is normally acceptable; however, with germanium transistors, especially when operated at high temperatures, this factor can become a severe limitation to operation.

Another factor of great importance is that of component tolerances. As indicated above, the factor h_{FE} can seriously affect the ability of a multivibrator to operate. Similarly, the values of the other components will determine whether the input presents a negative resistance. Thus, in the final analysis, all components must be taken at certain limit values to insure operation under all combinations of tolerances and operating conditions. In the rigorous presentations to be found in more extensive treatments of the subject, this matter is emphasized, and equations are given in terms of maximum and minimum values of all pertinent parameters.

Figure 18.3 shows the circuit of a very simple astable multivibrator, in which the previous coupling resistors and emitter resistors have been omitted and the base resistors have been connected to the supply voltage. The frequency is approximately $1/2 \ln 2R_bC_k \simeq 0.8/R_bC_k$. One potential fault of this arrangement is that both transistors can saturate together, thus inhibiting startup. The circuit of Fig. 18.1 (with the added capacitors, of course) is much better in this respect, and reliable starting is assured, provided the components have the necessary values.

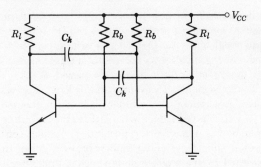

Fig. 18.3 Circuit of simple astable multivibrator.

18.3 The Monostable Multivibrator

The basic circuit of Fig. 18.1 can be converted to a monostable arrangement by simple modification, as illustrated in Fig. 18.4. The two emitters now share a common resistor R_f; however, an additional resistor R_{E1} is now inserted in the first emitter lead. This dissymmetry

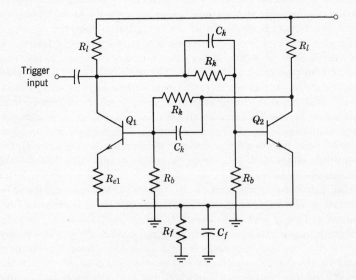

Fig. 18.4 Circuit of monostable multivibrator.

results in transistor Q_2 biasing transistor Q_1 into cutoff normally, Q_2 being held at the edge of saturation. On imposition of a negative pulse at the trigger input terminals Q_2 is driven toward cutoff. Its collector voltage rises and the resultant positive-going signal on the base of Q_1 produces a regenerative action that finally reverses the state of the two transistors. After an interval, determined by the decay of the charge built up on capacitors C_k, the regenerative action will take place in reverse and the circuit will revert to its original state. Thus, the effect of the input pulse has been to produce an output pulse of constant width and height, independent of the input excitation, provided only that it is sufficient to start the regenerative action. By proper choice of transistors and associated circuit components, the output pulse can be made to have duration ranging from tens of nanoseconds to seconds, amplitude on the order of 5–10 v. Such circuits are of considerable value in such applications as pulse counting, frequency measurement, and other such uses where uniformity of pulse shape is important.

18.4 The Bistable Multivibrator

This device is the work horse of the computer industry and, in one form, is the well-known "flip-flop." Basically it is the same circuit as that shown in Fig. 18.4, with the resistor R_{E1} removed. Thus the circuit is completely symmetrical and has two stable states, provided the component values and the transistor parameters can give the necessary negative resistance. In the analysis of the bistable multivibrator as well as of the monostable one, the same procedure applies as for the astable, as described in Section 18.2. One of the emitter leads is opened and a voltage source is inserted. The curve of emitter current versus voltage is obtained and, if it shows a negative-resistance characteristic, oscillation is possible. In these two cases, however, it will be found that a load line of zero resistance, i.e., a short-circuit, passing through the origin will intersect the characteristic curve in only one place for the monostable operation, and in three for the bistable one. In the latter case, the middle one will be unstable, hence the two stable states will be determined by the intersections in the positive-resistance regions.

Two modes of bistable operation are commonly used. In one, incoming pulses are applied to the same terminal and successively reverse the state of the bistable. In effect the bistable becomes a count-by-two flip-flop, and is used in this fashion in counters. In the other mode of operation inputs are applied to separate terminals, one being a "set" terminal, the other the "reset." Reversal of state only

184   TRANSISTOR APPLICATIONS

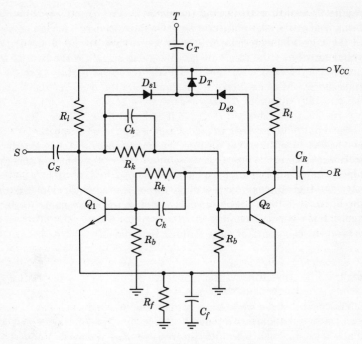

Fig. 18.5   Circuit of bistable multivibrator.

occurs when pulses are fed to the two terminals alternately. Once the bistable has been "set," additional signals applied to that terminal will produce no effect, and change of state will only occur on imposition of an input on the "reset" terminal.

These features are illustrated on Fig. 18.5, which shows the same basic bistable as in Fig. 18.4, but with the addition of steering diodes and the three inputs for trigger, set and reset inputs. The steering diodes D_{s1} and D_{s2} receive either a forward or reverse bias, depending on which transistor is cut off and which operative. Thus, an incoming pulse on the trigger terminal T will be steered to the proper side of the bistable to turn the off transistor on, and vice versa. This also reverses the states of the steering diodes, so that the next incoming pulse is steered to the opposite side. The diode D_T discharges the capacitor C_T and permits high-speed operation. The other two inputs can only produce one type of response. If Q_2 is on, a negative pulse on the S terminal will turn it off, and it will remain in that state until a negative pulse is applied to the R terminal (it is assumed in the above that all the signals are in the form of negative pulses without appreciable overshoots). This form of bistable is used very extensively in such computer components as registers.

18.5 The Schmitt Trigger

This device, illustrated in Fig. 18.6 is, in effect, a d-c bistable, changing state suddenly when the input level reaches a certain threshold, then shifting back when it drops down below a reset level. Normally, in the absence of a voltage on the input terminal, Q_1 is cut off by the emitter current of Q_2 through the 560-ohm resistor. When the input signal level becomes sufficient to begin conduction in Q_1, a regenerative action occurs and the states of the two transistors quickly reverse. They will remain in this new state until the input level drops sufficiently to cut Q_1 off again. With the values shown in Fig. 18.6, Q_1 always conducts if the input is more negative than -5 v. Q_2 always conducts if the input is less negative than -2 v. The output voltage change is greater than 8 v. The circuit can follow input changes up to a frequency of 500 kc; thus it can also be used to square a sine wave input.

The foregoing has described only a few basic types of the great number of variations which are used today. Some are simpler than those described, e.g., the direct-coupled types of multivibrators; some are

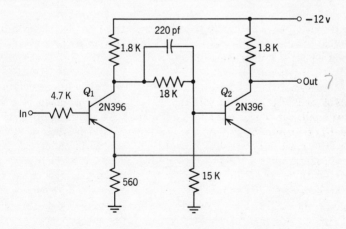

Fig. 18.6 Schmitt trigger.

more complex, e.g., using complementary combinations; some saturate, others use special circuit arrangements to avoid saturation and thereby increase counting speed. All use the same basic principles in one shape or another, and the foregoing should help to understand basically how they work.

REFERENCES

1. Greiner, R. A., *Semiconductor Devices and Applications*, McGraw-Hill, New York, 1961, pp. 386–409.
2. Hurley, R. B., *Transistor Logic Circuits*, Wiley, New York, 1961.
3. Schwartz, S. (Ed.), *Selected Semiconductor Circuits Handbook*, Wiley, New York, 1960, pp. 6-1–6-64.
4. Shea, R. F. (Ed.), *Transistor Circuit Engineering*, Wiley, New York, 1957, pp. 241–264, 324–356.
5. General Electric Co., *Transistor Manual*, 6th edition, pp. 155–174.
6. Texas Instruments Corp., *Transistor Circuit Design*, McGraw-Hill, New York, 1963, pp. 369–381.

19

Transient Response

19.1 Introduction

In previous chapters, the analysis of circuit behavior was based on steady-state conditions. In designing switching circuits and in other applications where wave shape, rise time, fall time, etc., are important, it becomes necessary to analyze circuits with respect to their response to various impulses, e.g., step functions; thus we must transfer our analysis from a frequency domain to a time domain, using, for example, the Laplace transforms.

When the amplitude excursions are small enough so that operation remains within the active region of the transistor characteristics, we can use the small-signal relationships developed earlier in the book, transforming them to the time domain. In many applications, however, operation ranges from cutoff to saturation, and is characterized by the term large-signal. Under these conditions, the small-signal techniques no longer apply and must be replaced by more suitable methods. Earlier texts treated large-signal transient response by means of the Ebers-Moll relationships, involving the normal and inverse amplifications and cutoff frequencies of the transistor. Later methods employ the use of charge control parameters, and this method will be followed in this chapter. The two basic methods are compatible; however, few specification sheets supply the inverse parameters and progressively more are now supplying charge control parameters.

19.2 Small-Signal Transients

As mentioned above, as long as the transistor is neither cut off nor saturated during its operation, the usual small-signal a-c parameters may be used to predict transient performance, although there may be considerable variation in the values of these parameters with excursion of operating point.

In order to apply the equations for small-signal parameters to this purpose, they are converted to the operational form, using the symbol s to indicate complex frequency, in this case the equivalent of $j\omega$. For example, the expression for h_{fb} as a function of frequency is:

$$h_{fb} \simeq \frac{(h_{fb})_0}{1 + j\omega/\omega_{ab}} \tag{19.1}$$

Using s in place of $j\omega$, this can be written as

$$h_{fb} \simeq \frac{(h_{fb})_0}{(1 + s/\omega_{ab})} \tag{19.2}$$

The approximate equations relating collector current to emitter current and emitter-base voltage to emitter current can be similarly written:

$$\frac{i_C(s)}{i_E(s)} \simeq \frac{(h_{fb})_0}{1 + s/\omega_{ab}} \tag{19.3}$$

and

$$\frac{v_{EB}(s)}{i_E(s)} = \frac{r_\epsilon + r_b'[1 + (h_{fb})_0 + s/\omega_{ab}]}{1 + s/\omega_{ab}} \tag{19.4}$$

Note the use of lower-case symbols with capital subscripts to indicate absolute values of time-varying quantities. The symbol $i_C(s)$, for example, can be read as the instantaneous value of the collector current as a function of frequency. Similarly, $i_C(t)$ is the same current, now expressed as a function of time.

Equations 19.3 and 19.4 will be recognized as the current amplification and input impedance (output short-circuited) respectively, with s used in place of $j\omega$. This implies that we are assuming that the effective load is low enough so that the transistor can be assumed to be working into an effective short-circuit. This is a reasonable assumption in transient analysis, as was also the case in the analysis of wide-band amplifiers, as low-impedance loads are normally required to obtain adequate bandwidth and, hence, frequency response.

To obtain now the transient response of a transistor circuit to some input function, we must express this function in operational form, insert it in the pertinent equations, obtain the resultant equations for output as functions of frequency, then convert back to the time domain.

TRANSIENT RESPONSE

As an aid in performing these operations, Table 19.1 gives the most commonly used relationships between time functions and their Laplace transforms.

As an example of the application of this technique, let us determine the transient collector current in response to a step in emitter current. Using eq. 19.3 and substituting $\Delta i_E/s$ for the step in emitter current, we obtain:

$$i_C(s) \simeq \frac{i_E(s)(h_{fb})_0}{1 + s/\omega_{ab}}$$

$$\simeq \frac{\Delta i_E}{s}\left[\frac{(h_{fb})_0}{1 + s/\omega_{ab}}\right] \simeq (h_{fb})_0 \omega_{ab} \Delta i_E \left(\frac{1}{s^2 + \omega_{ab}s}\right) \quad (19.5)$$

Equation 19.5 will be recognized as having the Laplace operator $1/(s^2 + as)$, hence the time solution will contain the exponential $(-at)$ where a is ω_{ab}. The final time solution is, therefore:

$$i_C(t) \simeq (h_{fb})_0 \Delta i_E (1 - e^{-\omega_{ab}t}) \quad (19.6)$$

Similarly, the response of i_C to a step change in v_{EB} can be obtained by using eqs. 19.3 and 19.4:

$$\frac{i_C(s)}{v_{EB}(s)} = \frac{i_C(s)}{i_E(s)} \frac{i_E(s)}{v_{EB}(s)} \simeq \frac{(h_{fb})_0}{r_\epsilon + r_b'[1 + (h_{fb})_0 + s/\omega_{ab}]} \quad (19.7)$$

Table 19.1   Common Laplace Transform Relationships

Time Function	Frequency Function
k	$\dfrac{k}{s}$
e^{-at}	$\dfrac{1}{s+a}$ or $\dfrac{s}{s^2 + as}$
$\dfrac{1 - e^{-at}}{a}$	$\dfrac{1}{s^2 + as}$
$\dfrac{ae^{-at} - be^{-bt}}{a - b}$	$\dfrac{s}{(s+a)(s+b)}$

Substituting $\Delta v_{EB}/s$ for $v_{EB}(s)$ and solving, we obtain, for the frequency response:

$$i_C(s) \simeq \frac{(h_{fb})_0 \omega_{ab} \Delta v_{EB}}{r_b'} \left\{ \frac{1}{s^2 + \omega_{ab}[1 + (h_{fb})_0 + r_\epsilon/r_b']s} \right\} \quad (19.8)$$

Again, the operational factor is of the form $1/(s^2 + as)$, thus the solution is:

$$i_C(t) \simeq \frac{(h_{fb})_0 \Delta v_{EB}}{r_\epsilon + r_b'[1 + (h_{fb})_0]} \left\{ 1 - \exp\left[1 + (h_{fb})_0 + \frac{r_\epsilon}{r_b'}\right](-\omega_{ab}t) \right\} \quad (19.9)$$

Note: if the response to a step in generator voltage v_G is desired, the same equation is used with $R_G + r_\epsilon$ substituted for r_ϵ.

19.3 Small-Signal Transient Response, Common-Base Configuration

The various transient responses to step inputs have been calculated as above and are summarized below. The parameter h_{fb} is used in these equations for simplicity [remembering that $h_{fb} = -h_{fe}(1 + h_{fe})$ and that $1 + h_{fb} = 1/(1 + h_{fe})$]. Furthermore, it is understood that this is the low-frequency specification value $(h_{fb})_0$—the added $(\ )_0$ being omitted to simplify these tabulations.

1. $i_C(t)$ in response to a step change Δi_E:

$$i_C(t) = h_{fb} \Delta i_E (1 - e^{-\omega_{ab}t}) \quad (19.10)$$

2. $i_C(t)$ in response to a step change Δv_{EB}:

$$i_C(t) = \frac{h_{fb} \Delta v_{EB}}{r_\epsilon + r_b'(1 + h_{fb})} \left\{ 1 - \exp\left[\left(1 + h_{fb} + \frac{r_\epsilon}{r_b'}\right)(-\omega_{ab}t)\right] \right\} \quad (19.11)$$

3. $v_{EB}(t)$ in response to a step Δi_E:

$$v_{EB}(t) = \Delta i_E [r_\epsilon + r_b'(1 + h_{fb}) - (r_\epsilon + h_{fb} r_b')e^{-\omega_{ab}t}] \quad (19.12)$$

4. $i_E(t)$ in response to a step Δv_{EB}:

$$i_E(t) = \frac{\Delta v_{EB}}{r_\epsilon + r_b'(1 + h_{fb})} \left\{ 1 + \left(\frac{r_\epsilon}{r_b'} + h_{fb}\right) \exp\left[\left(\frac{r_\epsilon}{r_b'} + 1 + h_{fb}\right)(-\omega_{ab}t)\right] \right\} \quad (19.13)$$

Add R_G to r_ϵ for response to Δv_G.

19.4 Small-Signal Transient Response, Common-Emitter Configuration

1. $i_C(t)$ in response to a step Δi_B:

$$i_C(t) = \frac{-h_{fb}}{1 + h_{fb}} \Delta i_B \{1 - \exp[(1 + h_{fb})(-\omega_{ab} t)]\} \quad (19.14)$$

2. $i_C(t)$ in response to a step Δv_{BE}:

$$i_C(t) = \frac{-h_{fb} \Delta v_{BE}}{r_e + r_b'(1 + h_{fb})} \left\{ 1 - \exp\left[\left(1 + h_{fb} + \frac{r_e}{r_b'}\right)(-\omega_{ab} t)\right] \right\} \quad (19.15)$$

3. $v_{BE}(t)$ in response to a step Δi_B:

$$v_{BE}(t) = \Delta i_B \left(r_b' + \frac{r_e}{1 + h_{fb}} \{1 - \exp[(1 + h_{fb})(-\omega_{ab} t)]\} \right) \quad (19.16)$$

4. $i_B(t)$ in response to a step Δv_{BE}:

$$i_B(t) = \frac{\Delta v_{BE}}{r_b'} \left\{ \frac{r_b'(1 + h_{fb}) + r_e \exp[(1 + h_{fb} + r_e/r_b')(-\omega_{ab} t)]}{r_e + r_b'(1 + h_{fb})} \right\} \quad (19.17)$$

For responses to step change in generator voltage Δv_G, substitute $R_G + r_b'$ for r_b'.

19.5 Small-Signal Transient Response, Common-Collector Configuration

1. $i_E(t)$ in response to a step Δi_B:

$$i_E(t) = \frac{\Delta i_B}{1 + h_{fb}} \{1 + h_{fb} \exp[(1 + h_{fb})(-\omega_{ab} t)]\} \quad (19.18)$$

2. $i_E(t)$ in response to a step Δv_{BC}:

$$i_E(t) = \Delta v_{BC} \left(\frac{1 + \dfrac{r_e + h_{fb} r_b'}{r_b' + R_l} \exp\left\{\left[\dfrac{r_b'(1 + h_{fb}) + r_e + R_l}{r_b' + R_l}\right](-\omega_{ab} t)\right\}}{r_e + r_b'(1 + h_{fb}) + R_l} \right) \quad (19.19)$$

3. $v_{BC}(t)$ in response to a step Δi_B:

$$v_{BC}(t) = \frac{\Delta i_B}{1 + h_{fb}} \{r_\epsilon + r_b'(1 + h_{fb}) + R_l - (r_\epsilon - h_{fb}R_l)$$
$$\exp\left[(1 + h_{fb})(-\omega_{ab}t)\right]\} \quad (19.20)$$

4. $i_B(t)$ in response to a step Δv_{BC}:

$$i_B(t) = \Delta v_{BC}$$
$$\left(\frac{1 + h_{fb} + \left[\dfrac{r_\epsilon - h_{fb}R_l}{r_b' + R_l}\right]\exp\left\{\left[\dfrac{r_b'(1 + h_{fb}) + r_\epsilon + R_l}{r_b' + R_l}\right](-\omega_{ab}t)\right\}}{r_\epsilon + r_b'(1 + h_{fb}) + R_l}\right)$$
$$(19.21)$$

For step changes in generator voltage Δv_G, substitute $R_G + r_b'$ for r_b'.

19.6 Large-Signal Transient Response; Charge Control Considerations

As mentioned in the introduction to this chapter, the charge control technique permits reasonably accurate calculations of the various characteristics of the output response to large-signal step inputs. This system is based on consideration of the charge requirements to compensate for the changes in depletion layer widths, to create a charge distribution pattern in the base, and to produce collector current flow.

Consider the elementary switching circuit of Fig. 19.1, which shows

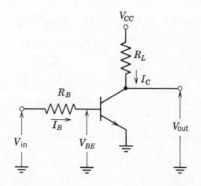

Fig. 19.1 Elementary switching circuit.

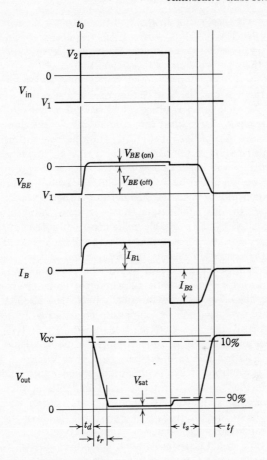

Fig. 19.2 Wave-forms of input voltage and current, and of output voltage, for step increase in input current.

an n-p-n transistor with collector supplied from supply V_{CC} through load resistor R_L and an input voltage V_{in} applied to the base through resistor R_B. Assume the input voltage to have the step-function shown in Fig. 19.2, starting from a negative voltage V_1 prior to switching, then jumping to a positive voltage V_2, remaining at that value for a while, then returning to its initial value V_1.

Prior to t_0 the transistor was cut off, no current was flowing in the collector circuit (neglecting I_{CBO}), and the output voltage V_{out} was at the supply level V_{CC}. Base current I_B was also zero, and the base voltage $V_{BE(\text{off})}$ equaled the input voltage V_1. Both emitter and collector junctions are wide under these conditions.

On imposition of the positive input voltage step, an input current I_{B1} flows into the base and the base voltage rises to a slightly positive value, $V_{BE(\text{on})}$. No change takes place yet in the output current or voltage because the input charge must first compensate for the charge required to forward bias the emitter junction (it is now narrowed), then to set up a charge distribution in the base. After a time t_d, the delay time, the collector current will begin to flow and the output voltage to drop. If there were no limitation on its amplitude, the collector current would increase exponentially in similar manner to that described for the small-signal case. In this application, however, the collector current is normally limited by the available value dictated by the supply battery and load resistor, or a maximum value of V_{CC}/R_L. This value of current is the saturation current and the resultant output voltage is the saturation voltage. Figure 19.3 shows the collector family for the 2N396A transistor—(a) showing the complete range of operation and (b) showing the operation at saturation. (*Note:* the 2N396A is a p-n-p unit, thus the voltages will have opposite polarities to those described above.) A load line corresponding to a resistor R_L of 1000 ohms has been drawn on these curves from a supply of 20 v. Thus maximum current is -20 ma and an input base current of 2 ma will just about produce this condition. V_{sat} is about 0.05 v under this condition.

Let us further assume, for the sake of illustration, that the input resistance R_B has a value of 20,000 ohms. The emitter family for this transistor shows that the voltage $V_{BE(\text{on})}$ will be about 0.4 v for the saturated condition described above. Thus an input voltage of slightly more than 40 v would be required to obtain a drive current I_{B1} of 2 ma, although an input of 1 ma is still enough to saturate essentially the transistor, producing an output voltage of only 0.1 v.

When the input signal is removed and reverts to its initial value V_1, the base voltage does not immediately return to its initial value but to a slightly lower value than during saturation, and remains at this value until the original charge distribution in the base is restored. The base current, therefore, reverses under the conditions shown and assumes a value I_{B2} approximately equal to V_1/R_B.

After a period t_s, the storage time, the collector current will again fall in accordance with small-signal theory until the initial conditions are restored. Thus, there are four times associated with the output pulse: t_d, the delay time, before any change occurs; t_r, the rise time, usually

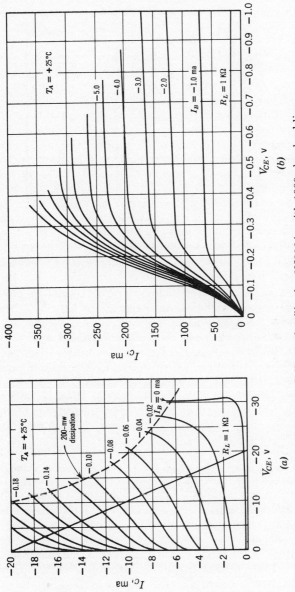

Fig. 19.3 Collector families for 2N396A, with 1000-ohm load line.

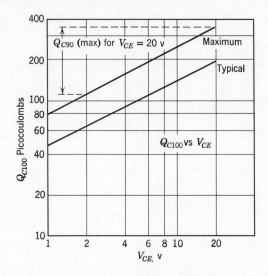

Fig. 19.4 Collector charge versus collector-emitter voltage, 2N396A.

measured between points at 10% and 90% of the final collector current excursion; t_s, the saturation time; and t_f, the fall time. These four times may be computed in the following manner.

Two lifetimes are given on the specification sheets, τ_a and τ_b, the lifetimes in the active region and in the saturated region respectively. Furthermore, an average value is given for the emitter capacitance, $\overline{C_{BE}}$. Finally, curves are given of the base and collector charges as functions of collector voltage and saturation current. Such a set of curves for the 2N396A are shown in Figs. 19.4 and 19.5. By using these curves and specification information, the four significant times can be calculated using the following equations:

$$t_d = \frac{|V_{BE(\text{off})}|\overline{C_{BE}}}{I_{B1}} + \left| \frac{Q_{B90} + Q_{C90}}{9(I_{B1} - 0.5I_{BS})} \right| \tag{19.22}$$

$$t_r = \left| \frac{8(Q_{B90} + Q_{C90})}{9(I_{B1} - 0.5I_{BS})} \right| \tag{19.23}$$

$$t_s = \tau_b \ln\left(\frac{I_{B1} - I_{B2}}{I_{BS} - I_{B2}}\right) + \left| \frac{(Q_{B100} - Q_{B90}) + (Q_{C100} - Q_{C90})}{-I_{B2} + 0.5I_{BS}} \right| \tag{19.24}$$

$$t_f = \left| \frac{8(Q_{B90} + Q_{C90})}{9(-I_{B2} + 0.5 I_{BS})} \right| \qquad (19.25)$$

Note: I_{B1} and I_{B2} are usually opposite in direction, hence their magnitudes will add, for example, in eq. 19.24. I_{BS} is the value of I_B necessary to produce I_{CS} or $I_{BS} = I_{CS}/h_{FE}$.

As an example of the use of the above equations, let us calculate the delay, rise, storage, and fall times for the following conditions:

2N396A transistor: $V_{CC} = -20$ v, $R_L = 1000$ ohms, $I_{CS} = -20 \times 10^{-3}$;

From spec sheets; $\tau_b = 0.65$ μs and $\overline{C_{BE}} = 12 \times 10^{-12}$;

Let $V_1 = 5$ v, $V_2 = -40$ v (polarities are reversed from Fig. 19.2 because of *p-n-p*).

Then

$$V_{BE(\text{off})} = 5.0 \quad \text{and} \quad I_{B1} = -2.0 \times 10^{-3}$$

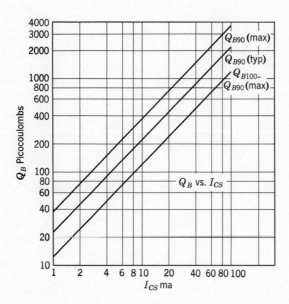

Fig. 19.5 Base charge versus saturation collector current, 2N396A.

From the curves $h_{FE} = 40$; therefore $I_{BS} = -0.5 \times 10^{-3}$. Using the curves for maximum values:

$$Q_{C_{100}} = 350 \times 10^{-12} \qquad Q_{B_{100}} = 970 \times 10^{-12}$$
$$Q_{C_{90}} = 240 \times 10^{-12} \qquad Q_{B_{90}} = 730 \times 10^{-12}$$

Substituting in eqs. 19.22–19.25, we obtain:

$$t_d = \frac{5.0 \times 12 \times 10^{-12}}{2.0 \times 10^{-3}} + \frac{730 \times 10^{-12} + 240 \times 10^{-12}}{9(2.0 \times 10^{-3} - 0.25 \times 10^{-3})}$$

$$= 30 \times 10^{-9} + 61.5 \times 10^{-9}$$

$$= 0.092 \ \mu s$$

$$t_r = \frac{8(970 \times 10^{-12})}{9(1.75 \times 10^{-3})} = 493 \times 10^{-9} = 0.493 \ \mu s$$

$$t_s = 0.65 \times 10^{-6} \ln \frac{2.25 \times 10^{-3}}{0.75 \times 10^{-3}} + \frac{110 \times 10^{-12} + 240 \times 10^{-12}}{0.25 \times 10^{-3} + 0.25 \times 10^{-3}}$$

$$= 0.71 \times 10^{-6} + 700 \times 10^{-9}$$

$$= 1.41 \ \mu s$$

$$t_f = \frac{8(970 \times 10^{-12})}{9(0.5 \times 10^{-3})} = 1720 \times 10^{-9} = 1.72 \ \mu s$$

REFERENCES

1. Greiner, R. A., *Semiconductor Devices and Applications*, McGraw-Hill, New York, 1961, pp. 344–366.
2. Shea, R. F. (Ed.), *Transistor Circuit Engineering*, Wiley, New York, 1957, pp. 299–317.
3. General Electric Co., *Transistor Manual*, 6th edition, pp. 65–100.

20

Negative-Resistance and Switching Devices

20.1 Introduction

Although the devices to be described in this chapter are not actually transistors in the strictest sense of the word, it seems appropriate to include them since they are active devices and do many of the things that transistors do, occasionally better. Included in this category are the unijunction transistor, the four-layer diode, and tunnel diodes and their relative, the backward diode.

20.2 The Unijunction Transistor

This device, originally called the double-base diode, is remotely related to the field-effect, or unipolar, transistor, one major distinction being that the input diode usually conducts during at least part of the device's operation. Figure 20.1 shows the convention adopted for the device and its elements. In effect, it is a filament of semiconductor material e.g., silicon, to which a rectifying junction is attached. Carriers injected by this junction modulate the conductivity of the bar and, in particular, of that portion between the emitter E and the lower base B_1. Figure 20.2 illustrates this by showing the unijunction transistor as a network of two resistors and a diode, the lower resistor being variable.

Figures 20.3 and 20.4 give the characteristics at room temperature of a typical unijunction transistor, the General Electric type 2N489. Figure 20.3 shows the typical N-shaped negative-resistance character-

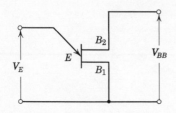
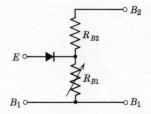

Fig. 20.1 Symbology for unijunction transistor.

Fig. 20.2 Equivalent circuit of unijunction transistor.

istic possessed by this device and its dependence on the applied interbase voltage. Figure 20.4 shows the relationship between interbase voltage, base-2 current and emitter current.

In the cutoff or standby condition, the emitter is back-biased with respect to the potential of that portion of the base adjacent to it. The interbase voltage sets up a voltage gradient along the length of the bar, with the result that the portion of the base adjacent to the emitter is at a voltage intermediate between the extremes of the interbase voltage. When the emitter is now biased in the forward direction, so that a portion

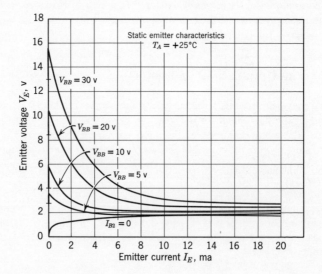

Fig. 20.3 Static emitter characteristics, 2N489 unijunction transistor.

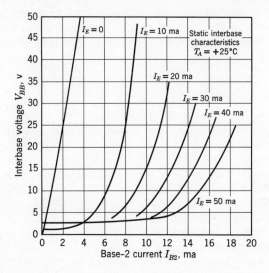

Fig. 20.4 Static interbase characteristics, 2N489 unijunction transistor.

of it becomes forward-biased with respect to the adjacent base, the emitter conducts and injects majority carriers into the base region below it. This reduces the resistance of the portion between E and B_1 and brings more of the emitter area into conduction, so that the process becomes cumulative and the device rapidly switches into the conducting mode. As a result of this action, the unijunction transistor has been used extensively as a generator of special waveforms, e.g., sawtooth, as a triggering device and for other similar uses.

The parameters of the unijunction transistor of major interest are:[*]

1. R_{BB}—*Interbase resistance.* This is the resistance measured between base 1 and base 2 with the emitter open. It increases with temperature at about $0.8\%/°C$.
2. η—*Intrinsic standoff ratio.* A function of peak-point emitter voltage, interbase voltage, and junction temperature. It is given by the equation

$$V_P = \eta V_{BB} + \frac{200}{T_J} \qquad (T_J \text{ in } °K)$$

[*] See Ref. 8, end of chapter.

202 TRANSISTOR APPLICATIONS

3. I_P—*Peak point current.* The emitter current at the peak points of the curves of Fig. 20.3. Although this figure seems to show the peaks occurring at zero current it can actually have a value as high as 20 μa.
4. V_P—*Peak point emitter voltage.* The voltage at the peaks of the curves of Fig. 20.3.
5. $V_{E(\text{sat})}$—*Emitter saturation voltage.* The forward drop from the emitter to base 1 in the saturation region.
6. $I_{B2(\text{mod})}$—*Interbase modulated current.* The base-2 current with the emitter saturated.
7. I_{EO}—*emitter reverse current.* The current between emitter and base 2, with base 1 open. This current is somewhat analogous to the I_{CBO} of a transistor.
8. V_V—*valley voltage.* The emitter voltage at the valley point. This voltage increases with interbase voltage, decreases with the addition of resistance in series with base 2, and increases with resistance in series with base 1.
9. I_V—*valley current.* The emitter current at the valley point. It increases with interbase voltage and decreases with introduction of resistance into either base lead.

Typical values of the above parameters for the 2N489 are:

R_{BB} 4700–6800 ohms at room temperature.
η 0.51–0.62 ($V_{BB} = 10$ v).
I_P 20 μa maximum ($V_{BB} = 25$ v).
V_P see Fig. 20.3. It is given by $\eta V_{BB} + 200/T_J$.
$V_{E(\text{sat})}$ 5 volts maximum ($I_E = 50$ ma, $V_{BB} = 10$ v).
$I_{B2(\text{mod})}$ 6.8–22 ma ($I_E = 50$ ma, $V_{BB} = 10$ v).
I_{EO} 20 μa maximum ($V_{B2E} = 30$ v, $I_{B1} = 0$).
I_V 8 ma minimum ($V_{BB} = 20$ v, $R_{B2} = 100$ ohms).

Figure 20.5 illustrates an elementary relaxation oscillator using the unijunction transistor. The capacitor C charges through the resistor R_1 until it reaches the point where the emitter becomes forward-biased, whereupon the emitter-base region assumes very low resistance and discharges C. The conditions for oscillation are:

$$\frac{V_{BB} - V_P}{R_1} > I_P \qquad (V_{BB} - V_V) < I_V$$

R_1 can vary from about 2000 ohms to 2 megohms. R_2 is used for temperature compensation and its value may be calculated from the equation:

$$R_2 \simeq \frac{0.40 R_{BB}}{\eta V_{BB}}$$

The maximum and minimum values of emitter voltage are V_P and approximately $0.5 V_{E(\text{sat})}$ respectively. The frequency of oscillation is given by:

$$f = \left[R_1 C \ln\left(\frac{1}{1-\eta}\right) \right]^{-1} \quad (20.1)$$

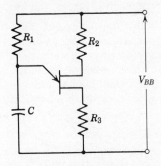

Fig. 20.5 Relaxation oscillator utilizing unijunction transistor.

The voltage at the emitter is approximately a sawtooth, although the rising portion of the waveform is not extremely linear. The output from the base-2 terminal is a series of positive pulses, from the base-1 terminal, negative pulses. These may be used, for example, to trigger other higher power circuits, e.g., those employing silicon controlled rectifiers.

Figure 20.6 illustrates a more linear saw-tooth generator, with a

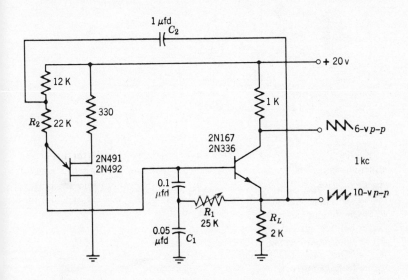

Fig. 20.6 Linear sawtooth generator utilizing unijunction transistor.

possible linearity better than 0.3%. The n-p-n transistor serves as a buffer and bootstraps the oscillator emitter through capacitor C_2 and resistor R_2. R_1 and C_1 give integrator-type feedback which compensates for the loading of the output stage.

Many other applications are possible for the unijunction transistor. Additional examples are: voltage sensing circuits; staircase wave generator; time delay relay; multivibrators; hybrid timing circuits. These are described in Ref. 8, end of chapter.

20.3 The Four-Layer Diode

This device consists of four layers, as contrasted to the three of the usual transistor. This can be considered as a combination of a p-n-p transistor and an n-p-n transistor, with the n-type base of the first connected to the emitter of the second, and the p-type base of the second connected to the collector of the first, internally. The external leads

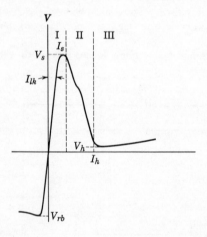

Fig. 20.7 Characteristic curve of four-layer diode, with terms and symbols.
V_s = switching voltage
I_s = switching current
I_h = holding current
V_h = holding voltage
R_{on} = "On" resistance (the slope of the VI curve measured at currents $> I_h$)
I_{lk} = leakage current
V_{rb} = Reverse breakover (avalanche) voltage

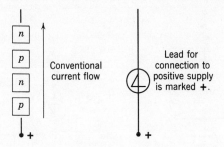

Fig. 20.8 Symbol for four-layer diode.

are the emitter of the *p-n-p* and the collector of the *n-p-n*. In solving this composite transistor circuit, the emitter current of the *p-n-p* transistor (which, of course, is also the collector current of the *n-p-n*) is given as:

$$I_{E(P)} = \frac{I_{CBO(N)} + I_{CBO(P)}}{1 - (\alpha_N + \alpha_P)} \qquad (20.2)$$

where α_N and α_P are the d-c current amplifications of the two units. This indicates that this simple two-terminal device can become unstable if the sum of the amplifications becomes unity. In addition, by inserting a control current to the base of the *n-p-n* transistor, the relative values of the two amplifications can be changed. The basic two-terminal device is the four-layer diode and the three-terminal device becomes the controlled switch or controlled rectifier.

Figure 20.7 shows the characteristics of one commercial four-layer diode, the Shockley, with the symbology explained in the attached table. Figure 20.8 illustrates the current convention and the symbol for the device employed by Shockley.

Referring to Fig. 20.7, there are two stable states—the "off" state, or high-resistance state, in region I, and the "on," or low-resistance state, in region III. To turn the device on, the voltage across the terminals must exceed the switching voltage V_s. It is turned off by reducing the current below the holding current I_h. Figure 20.7 is not linear, but is expanded to illustrate the operation of the device. The current at the switching point is typically several microamperes. When it has switched on, the current is limited mainly by resistance in the external circuit, since the diode has a dynamic resistance on the order of a few

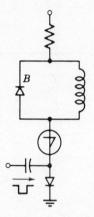

Fig. 20.9 Four-layer diode relay drive.

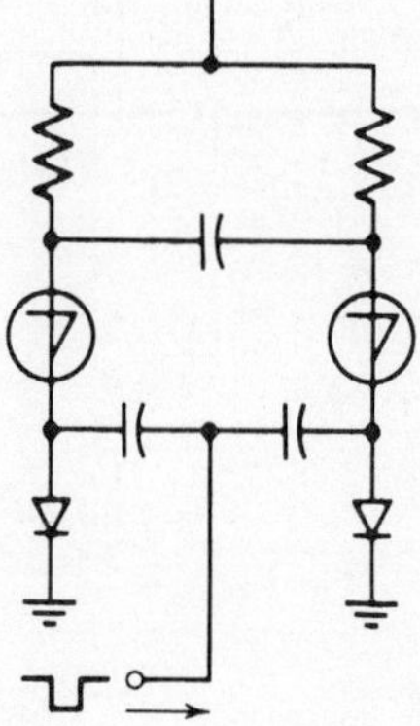

Fig. 20.10 Four-layer diode flip-flop circuit.

ohms and a voltage drop of about 1 v. As long as the current exceeds the holding current I_h, the device will remain in this high-conductivity state. When the current drops below this value, it will switch back to the "off" state. Devices are available with switching voltages from 20 to 200 v and holding currents from 1 to 50 ma.

Figures 20.9 through 20.13 illustrate a number of applications of this simple device. Details on these and many more may be obtained from the Shockley Transistor Unit of Clevite Transistor, Palo Alto, Cal.

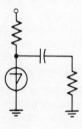
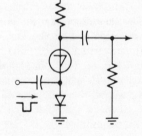

Fig. 20.11 Four-layer diode sawtooth oscillator.

Fig. 20.12 Four-layer diode pulse generator.

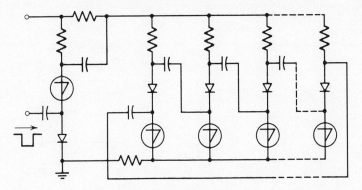

Fig. 20.13 Four-layer ring counter.

20.4 Tunnel Diodes

The tunnel diode is a unique two-terminal device possessing a pronounced negative conductance characteristic, which has demonstrated its applicability in a large number of roles, both for switching and linear operation.

Figure 20.14 shows the basic characteristic of the tunnel diode. It is similar in shape to the characteristic of the unijunction transistor, although with a more pronounced valley point. There is one other major difference, however, in that the two axes are reversed, hence the tunnel diode can more properly be called a negative-conductance device. Figure 20.14 also shows the small-signal equivalent circuit of the device, and the symbol used by G.E. for the tunnel diode.

The characteristic has a relatively high resistance in the third-quadrant section and in the first quadrant up to the peak point V_P, I_P. After traversing the negative-conductance region, it reaches the valley point, V_V, I_V, then turns up again. The other points indicated on this figure are the negative voltage V_R corresponding to a negative current equal and opposite to I_P and the forward voltage V_{FP} at the current I_P. Typical values of these currents and voltages, for the 1N2939 diode, for example, are:

I_P 1.0 ma V_P 65 mv V_R 30 mv max,
I_V 0.1 ma V_V 350 mv V_{FP} 500 mv

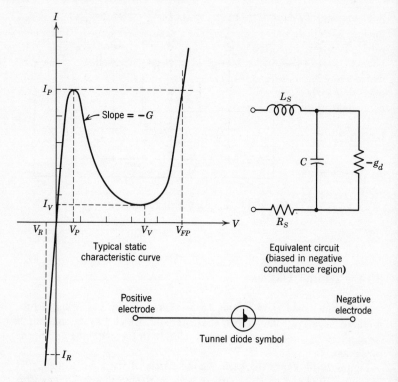

Fig. 20.14 Typical tunnel diode characteristic, equivalent circuit, and symbol.

Typical dynamic characteristics for the same tunnel diode are:

Total series inductance $L_S = 4 \times 10^{-9}$ h (depends on lead length)
Total series resistance $R_S = 1.5$ ohms
Valley point terminal capacitance $C = 5$ pf
Max. negative terminal conductance $-G = 6.6 \times 10^{-3}$ mhos
Resistive cutoff frequency $f_{ro} = 2.2$ gc
Self-resonant frequency $f_{xo} = 1.1$ gc
Frequency of oscillation $f_{osc} = 1.1$ gc

The resistive cutoff frequency is that frequency at which the real part of the diode impedance measured at its terminals goes to zero. The diode cannot amplify above this frequency. The self-resonant frequency is that at which the imaginary part of the impedance goes to zero. Both of these frequencies are affected by external circuit components and, therefore, the highest operating frequency is circuit-dependent. In a transistor package, lead inductance usually restricts operation to below 1 gc; however, microstrip or microwave packaging can extend the frequency to 5 gc or higher.

The equation for the resistive and self-resonant cutoff frequencies can be derived from the equivalent circuit of Fig. 20.14:

$$f_{ro} = \frac{|g_d|}{2\pi C} \sqrt{\frac{1}{R_S|g_d|} - 1} \qquad (20.3)$$

$$f_{xo} = \frac{1}{2\pi} \sqrt{\frac{1}{L_S C} - \left(\frac{g_d}{C}\right)^2} \qquad (20.4)$$

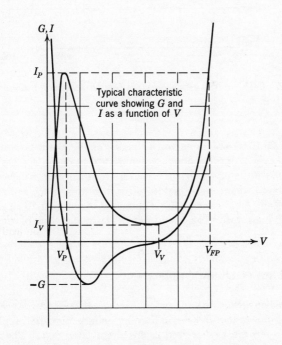

Fig. 20.15 Current and conductance versus voltage for tunnel diode.

210 TRANSISTOR APPLICATIONS

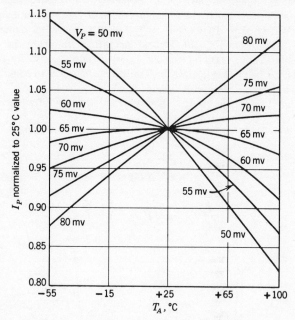

Fig. 20.16 Peak current versus ambient temperature as a function of peak voltage.

Figure 20.15 shows an enlarged version of the characteristic of Fig. 20.14, also a plot of the conductance, obtained from the slope of the first curve. It will be noted that the negative conductance reaches a negative maximum about in the middle of the downward slope and is reasonably constant for some extent. Figures 20.16 through 20.18 show the effects of temperature on the principle characteristics of the tunnel diode.

20.5 Tunnel Diode Applications

Obviously a device with the pronounced negative-conductance characteristic of the tunnel diode has literally endless potential applications and only a few can be described in this abbreviated text. The reader is referred to the excellent manuals listed in the references at the end of the chapter for detailed information on the wide variety of uses to which this device has been applied. These applications range from linear ones, such as sinusoidal oscillators and amplifiers, to switching and logic applications, such as flip-flops, monostable and astable multivibrators, staircase wave generators, trigger circuits, threshold detectors, and a host of similar circuits.

Figure 20.19 shows the circuit of a 100-Mc tunnel diode amplifier. Measured results were 32-db gain with 20 Mc bandwidth. Complete details of the design of this amplifier are given in Ref. 9, end of chapter. Similar tunnel diode amplifiers have been designed, extending operation into the gigacycle region by means of strip-line techniques.

Figure 20.20 shows the circuit of a 100-kc oscillator. Applying the same basic principle, tunnel diode oscillators have been made to operate at frequencies as high as 6 gc, using microwave cavity construction to house the diode.

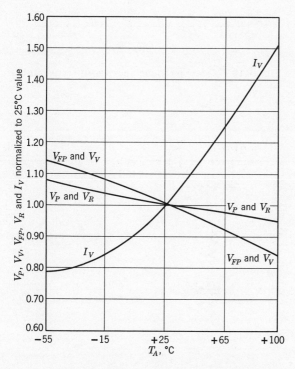

Fig. 20.17 Temperature characteristics of tunnel diode.

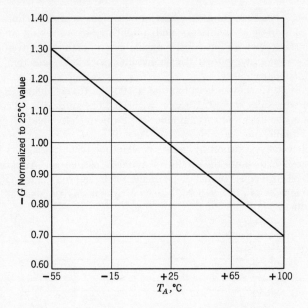

Fig. 20.18 Negative conductance versus temperature.

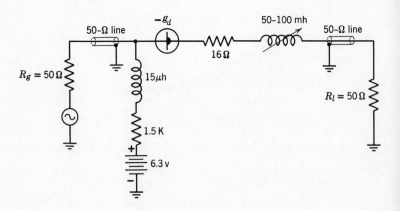

Fig. 20.19 Circuit of 100-Mc tunnel diode amplifier.

NEGATIVE-RESISTANCE AND SWITCHING DEVICES 213

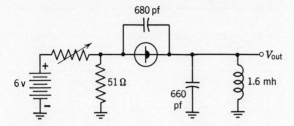

Fig. 20.20 Circuit of 100-kc tunnel diode oscillator.

The above two circuits have illustrated the linear applications of tunnel diodes. Of possibly even greater interest are the various switching applications, as these permit utilization of these devices in a great number of computer applications, for threshold detectors, and for the generation of nonsinusoidal waveforms, e.g., square waves.

Referring again to the basic tunnel diode characteristic of Fig. 20.14, it is evident that the device has three possible modes of operation as a switching device, depending on the location of the load line. If the

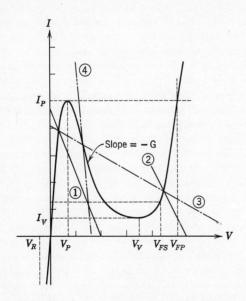

Fig. 20.21 Load lines for various modes of operation.

214   TRANSISTOR APPLICATIONS

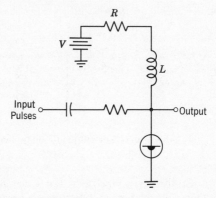

Fig. 20.22   Components of tunnel diode multivibrator.

load line intersects the curve in only one point, in the negative-conductance region, it is potentially a free-running relaxation oscillator, in similar manner to the astable multivibrator described in Chapter 18. If the single intersection is either in the linear portion to the left of the peak point or to the right of the valley point, the circuit will have one stable state and may be used as a monostable multivibrator. If the load line intersects in two stable points, the operation as a bistable multivibrator is possible. These three conditions are illustrated in Fig. 20.21. Load lines 1 and 2 intersect in positive-conductance regions only, hence represent monostable operation; load line 3 intersects in two stable regions (the third intersection is unstable), hence, is bistable; and line 4 represents a potential astable oscillator.

Figure 20.22 illustrates the components required to obtain these modes of operation. The d-c load line is set by the value of the resistor R, and the inductor L provides the energy storage necessary to permit oscillation or switching. If astable operation is wanted, the input connections are omitted. In the monostable mode, if the stable operating point is to the left of the peak point, positive input pulses will drive the current up to the peak, whereupon the circuit will switch along a constant-current line to the opposite point on the high-voltage portion of the characteristic. There will then be a relatively slow discharge down to the valley point, then a rapid switchover to the low-voltage portion, and finally the original point will be re-established. Had the stable point been on the high-voltage portion, similar action would have been initiated by negative input pulses.

In the bistable connection, as shown, positive input pulses will trigger the circuit from a low-voltage stable point to a high-voltage one. A negative pulse will then be required to switch it back to the original state. In this arrangement, the device acts like the set-reset multivibrator of Chapter 18. Two tunnel diodes can be arranged in a flip-flop, whereby similar input pulses will alternately switch the circuit back and forth between the two stable states. This arrangement will be described later in the chapter.

Hybrid arrangements using tunnel diodes and transistors have several interesting features. One such circuit is shown in Fig. 20.23. The transistor is normally cut off, since the tunnel diode is in the low-voltage state until triggered. When the diode switches to the high-voltage state, the transistor is turned on and an output pulse occurs at its collector. Figure 20.24 shows a further elaboration of this circuit, wherein the diode current is also controlled by the transistor. With the diode in its low-voltage state and the transistor cut off, the diode is biased just below its peak current by the current through the three resistors to the 5.5-v source. An input pulse as low as 70 mv can trip the circuit, as long as the pulse widths are between 0.5 and 5 μs. When the transistor is turned on, the collector voltage will fall and the diode will be reset at its low-voltage state.

Figure 20.25 illustrates the use of a string of tunnel diodes for frequency division, in this case, 5:1. The string of diodes is biased from a source supplying a current below the peak current rating of the diodes, thus initially all the diodes are in their low-voltage states. The bottom diode is chosen to have the highest peak current of the five. Each time

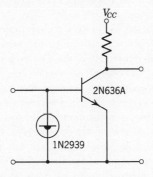

Fig. 20.23 Hybrid tunnel diode—transistor combination.

216 TRANSISTOR APPLICATIONS

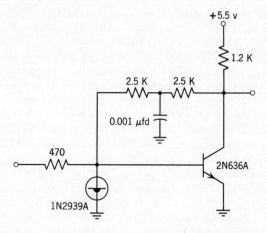

Fig. 20.24 Control of tunnel diode current by transistor.

a positive pulse occurs at the input, one of the top four diodes will switch to its high-voltage state until, on the fifth pulse, the bottom diode switches. This turns on the transistor and thus drops the voltage across the string of diodes so that they all reset to the original state. The output from the collector of the transistor will have the staircase wave shape shown, that across the lowest diode will be one pulse after five input pulses. The capacitor across the bottom diode and the inductor in series with the transistor base delay the pulse so that complete switching occurs. The operating frequency is mainly limited by the switching speed of the reset transistor.

Figure 20.26 shows the use of two tunnel diodes in a flip-flop. Here it is desired that successive input pulses reverse the state of the diodes so that the circuit is effectively a divide-by-two circuit. The supply voltage is such that only one of the diodes can be in the high-voltage state at a time, and the difference between the two tunnel diode currents flows through the inductance. When a positive trigger pulse turns on the diode which was in the low-voltage state, the voltage induced in the inductance (because of the decreasing current through it) resets the other tunnel diode to the low-voltage state. The next input pulse reverses the operation and resets conditions to their initial state. Each pair of trigger pulses completes one switching cycle. This basic flip-flop can be used in a counter to perform the usual divide-by-2, 4, 8, 16, etc., operation, up to a frequency on the order of 10 Mc.

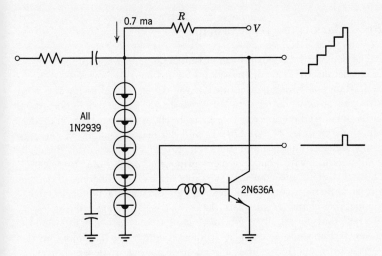

Fig. 20.25 Tunnel diode frequency divider.

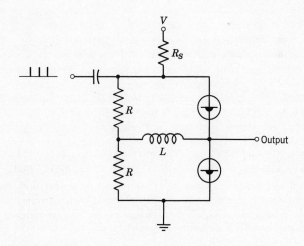

Fig. 20.26 Tunnel diode flip-flop.

20.6 Backward (Back) Diodes

The backward diode is a form of tunnel diode in which other features than the negative conductance are of primary importance. Figure 20.27 shows the principal features of the device (the drawing is not to scale and is only intended to illustrate characteristics of importance). In the forward direction the voltage is much less than for silicon diodes, being on the order of 120–170 mv at 1–25 ma, depending on the particular diode. Furthermore, the temperature stability in this direction is outstanding, with almost no perceptible shift from $-85°$ to $+200°C$. Thus, this diode provides a very stable low-voltage reference. In the reverse direction, some backward diodes evidence a minor negative-conductance characteristic, with the reverse peak current often as low as 10 μa, so that potentially they can be used as very low-current switches, in much the same manner as tunnel diodes. Furthermore, in the reverse direction, they exhibit a breakdown characteristic which permits their use as clippers and limiters. They are also used extensively as unidirectional coupling between tunnel diode stages.

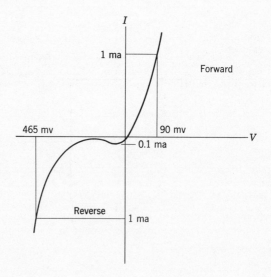

Fig. 20.27 Characteristic of backward diode.

REFERENCES

1. Hauer, W. B., *A 4-mw 6-kmc Tunnel-Diode Oscillator*, Paper No. TA 7.2, presented at the International Solid-State Circuits Conference, Philadelphia, Pa. (Feb. 1962).
2. Komamiya, Y., *Microwave Logic Circuits Using Esaki Diodes*, Paper No. WPM 3.1, ibid (Feb. 1963).
3. Kruy, J. F., *High-Speed Arithmetic Unit Using Tunnel Diodes*, Paper No. WPM 3.3, ibid. (Feb. 1963).
4. Sear, B. E., *Kilomegacycle Tunnel Diode Logic Circuits*, Paper No. TM 5.2, ibid. (Feb. 1962).
5. Sear, B. E., J. S. Cubert, and W. F. Chow, *The Enhanced Tunnel-Diode Logic Circuit*, Paper No. WPM 3.2, ibid. (Feb. 1963).
6. Shea, R. F. (Ed.), *Transistor Circuit Engineering*, Wiley, New York, pp. 18, 247–248, 258–259.
7. Trambarulo, R., *A Low-Noise X-Band Esaki-Diode Amplifier*, Paper No. TA. 7.4, ibid (Feb. 1962).
8. General Electric Co., *Transistor Manual*, 6th edition, pp. 191–201.
9. General Electric Co., *Tunnel Diode Manual*.
10. Hoffman Electronics Corp., Tunnel and Uni-tunnel Diode Applications, *Electrical Design News* (March 1961).
11. Transitron Electronic Corp., *The Tunnel Diode Circuit Design Handbook*, AN-1359A.
12. Shockley Transistor Unit of Clevite Transistor, *Introduction to the Shockley 4-Layer Diode*, and catalog of Shockley 4-layer diodes.

Note: Abstracts of above papers presented at the 1962 and 1963 International Solid-State Circuits Conferences are contained in the respective digests of technical papers.

21

Field-Effect (Unipolar) Transistors

21.1 Introduction

The field-effect (unipolar) transistor is not a new device. Shockley described the basic design of this device in 1952,* and a number of variations of this basic principle have been described in the following years. The unijunction transistor, described in Chapter 20, is a related device, although differing considerably in its basic mode of operation. The great advances in manufacturing technology during the past several years, together with new developments in device fabrication, have combined to make the production of high-gain low-noise field-effect transistors feasible, and the device has therefore begun to enjoy a long-delayed popularity.

The field-effect transistor (commonly abbreviated FET) differs from conventional transistors in three basic characteristics: (1) The control electrode is never in the conducting state, as contrasted to the emitter of the usual transistor; (2) majority carriers play the dominant role in its operation rather than minority carriers; and (3) control is by means of constriction, or pinchoff, of the majority carrier flow. The terminology is different, following that proposed by Shockley, and the basic characteristics are quite different. It lends itself to many of the applications for which electron tubes have still shown advantages over transistors, e.g., high input impedance, and thus further increases the encroachment of solid-state devices into the area once possessed solely by tubes.

* See Ref. 1, end of chapter.

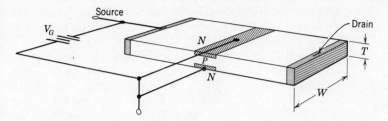

Fig. 21.1 Basic construction of the field-effect transistor.

21.2 Theory of Operation

A detailed theory of the operation of these devices is beyond the scope of this chapter, and the reader who is desirous of obtaining a detailed description is referred to the very comprehensive treatment contained in Ref. 6, end of chapter. The following material is largely extracted from this source, through the courtesy of Texas Instruments, and should suffice to provide an elementary understanding of the operation of the device and of some typical applications.

Figures 21.1 and 21.2 show, in very elementary fashion, the basic construction of the FET and its mode of operation. They also indicate the biasing conventions, symbol for the device, and terminology for the electrodes. In effect, the FET consists of a filament of semiconductor material with ohmic contacts (source and drain) at the ends, and a set

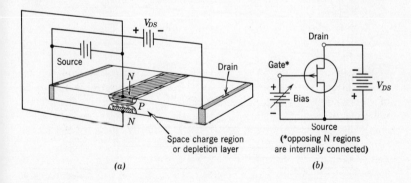

Fig. 21.2 (a) Illustration of operating principle of field-effect transistor; (b) Device symbol and voltage polarities.

222 TRANSISTOR APPLICATIONS

of opposing contacts which provide n-p junctions located between the above electrodes. Application of a voltage between the source and drain sets up an internal field along the length of the device, and a current will flow. Application of a voltage between the source and the gates produces a pair of depletion layers with a restriction of the carrier flow from source to drain resulting. As the gate voltage is varied, the pinchoff channel also varies, and the source-to-drain current is thus modulated. This is illustrated in Fig. 21.2, which also shows the wedge-shaped distribution of space charge between the gates, due to the IR drop along the channel. Thus, we see that the flow of current from the source to the drain is controlled by the transverse field created by the gate voltage, hence the name of the device.

21.3 Static Characteristics of the FET

Figure 21.3 shows typical static characteristics for the FET, in this case the 2N2497, and Fig. 21.4 illustrates some of the salient features of this set of curves. The meaning of the various currents and voltages is

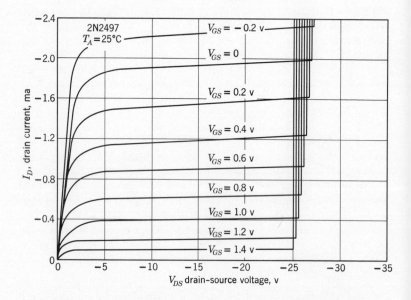

Fig. 21.3 Characteristic curves for the 2N2497 field-effect transistor.

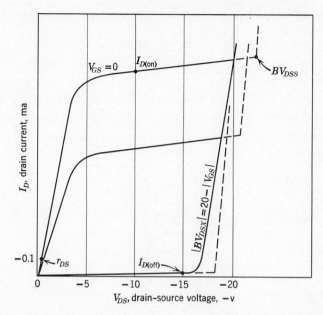

Fig. 21.4 Drain current versus drain-source voltage, illustrating important rating points.

indicated on Fig. 21.4. In this figure, BV_{DSX} denotes breakdown voltage, drain-to-source, where the subscript X denotes the value of BV_{DS} for a particular value of V_{GS}.

One value of particular importance is the so-called "on" current, $I_{D(on)}$, the drain current with zero bias, measured in the pinched-off region. This current is used quite frequently as a reference for drain current.

Figure 21.3 illustrates several interesting differences from the usual transistor characteristics. Most important is the use of gate voltage as the control variable, rather than emitter or base current. This also implies that mutual conductance rather than the familiar alpha or beta is the critical parameter, as is the case with electron tubes. For example, inspection of Fig. 21.3 indicates a g_m on the order of 1500 μmhos, a very respectable figure.

The values of current and voltage are on the same order of magnitude as with conventional transistors, although FETs are available with

224 TRANSISTOR APPLICATIONS

voltage ratings as high as 350 volts. The rapid breakdown is a characteristic of FETs.

21.4 Biasing the FET

Since FETs are similar in so many ways to electron tubes, it is natural to expect that they can be biased in similar manner. They also

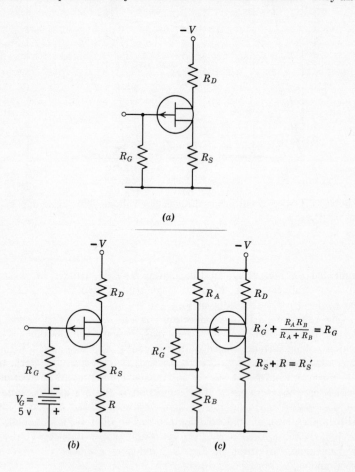

Fig. 21.5 Biasing the field-effect transistor.

possess some of the features of other transistors, such as variation with temperature, thus it is also to be expected that biasing would possess some of the problems found in conventional transistors. Such is the case, and generally FETs employ a combination of the two methods of biasing.

Figure 21.5 illustrates the three most common methods. Method a is similar to the familiar cathode bias method used with tubes. Method b increases the source resistance and inserts an extra bias in series with the gate, thereby improving stability and independence of variation in the parameter $I_{D(\text{on})}$. In method c the equivalent of the battery is achieved by means of the bleeder and voltage divider across the supply voltage. Table 21.1 gives typical values for these three methods, with design-center values shown for three types of FETs. The analytical technique for obtaining these values of bias resistors is not as simple as it appears, and the reader is referred to the above-referenced detailed descriptive material.

Table 21.1

FET Type	I_D Design Center, ma	R_S, kilohms	R_S', kilohms
2N2497	−0.5	2.00	12.0
2N2498	−1.25	1.00	5.0
2N2499	−4.00	0.53	1.8

21.5 Small-Signal A-C Parameters

As with all other active devices, or for that matter any network, the small-signal characteristics may be expressed in any of the forms presented earlier in the book. Reference 6 (end of chapter) presents a number of curves showing the variation of the various y parameters and their real and imaginary components with frequency. Figures 21.6 and 21.7 show the small-signal common-source input admittance and forward transfer admittance respectively as functions of frequency. It will be seen that the devices have useful properties up to several hundred megacycles.

21.6 Noise

One of the most promising features of FETs is their low noise figure, which, coupled with their high input impedance, makes them extremely useful for operation at very low signal levels. Figures 21.8 and 21.9 show the variations of spot noise with frequency, voltage, and current respectively. It will be seen that 0.5-db noise figure is achievable over quite considerable range of operation. Figure 21.10 shows the optimum noise figure and generator resistance versus frequency.

21.7 Typical Applications

Although the number of potential applications of FETs is almost limitless, a few typical examples will serve to indicate the various areas in which they have greatest utility.

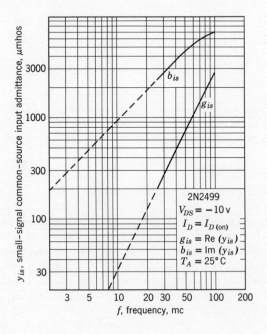

Fig. 21.6 Small-signal common-source input admittance versus frequency.

FIELD-EFFECT (UNIPOLAR) TRANSISTORS

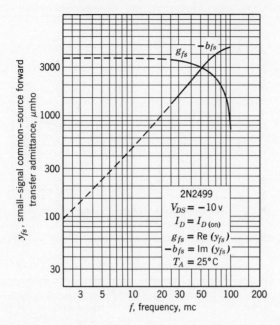

Fig. 21.7 Small-signal common-source forward transfer admittance versus frequency.

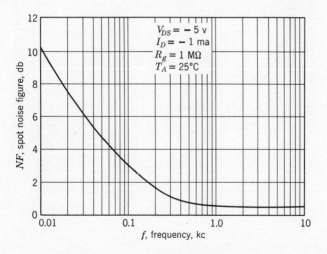

Fig. 21.8 Spot noise figure versus frequency.

228 TRANSISTOR APPLICATIONS

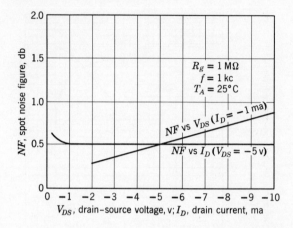

Fig. 21.9 Spot noise figure versus drain-source voltage and drain current.

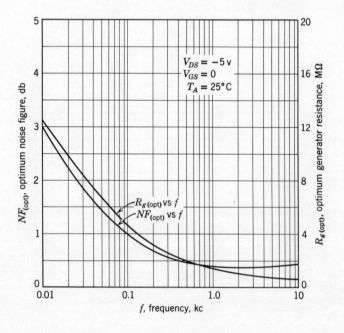

Fig. 21.10 Optimum noise figure and optimum generator resistance versus frequency.

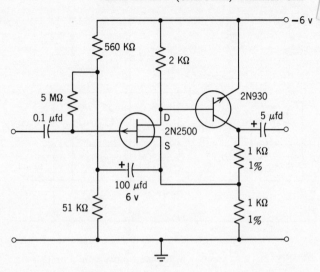

Fig. 21.11 Circuit of 6-db low-noise transducer amplifier.

The 6-db Low-Noise Transducer Amplifier. This amplifier, the circuit of which is shown in Fig. 21.11, was designed as a low-noise preamplifier for piezoelectric transducers. The input impedance varied from 180 megohms at 10 cps to 3 megohms at 10 kc, and the spot noise varied from 7 db to 1.2 db over the same range. The broad-band noise figure from 10 cps to 10 kc with a 200-kilohm generator resistance is 1.7 db; gain is 6 db. It will be noted that bootstrapping is used to reduce the loading effect of the bias resistors, thus permitting the above high values of input resistance.

Wide-Band Unity-Gain Amplifier. Figure 21.12 shows the circuit of an a-c amplifier in which the drain and the gate bias network of a source-follower input stage are driven by the output of the amplifier, thus reducing the effect of Miller capacitance and increasing the input resistance. The combined feedback effects produce an input impedance of about 100 megohms. Figure 21.13 shows the frequency response.

Figure 21.14 shows the circuit of a similar amplifier with response down to direct current and an input impedance on the order of 1000 megohms. The effect of gate-source capacitance is reduced by driving the source with the amplifier output. Effects of Miller capacitance are reduced by heavily loading the FET with a common-base transistor

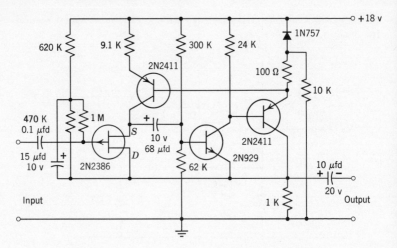

Fig. 21.12 Circuit of wide-band unity-gain amplifier.

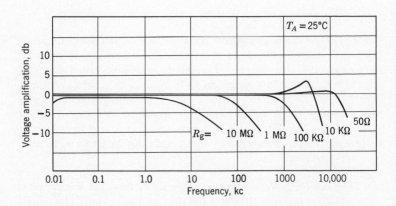

Fig. 21.13 Frequency response of amplifier of Fig. 21.12.

stage. This combination produces a large voltage amplification in the first two stages which is essential in obtaining gain accuracy. Two power supplies are required to obtain zero output for zero input.

Figure 21.15 shows the upper falloff frequencies for various generator resistances. By adjusting the trimmer capacitance shown in series with the input in Fig. 21.14, zero overshoot can be obtained at high frequencies.

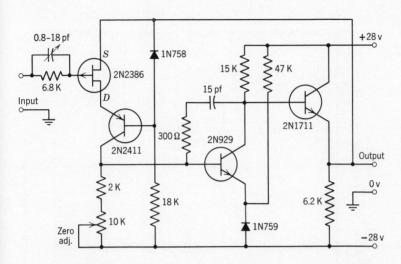

Fig. 21.14 Circuit of unity-gain high-input-impedance d-c amplifier.

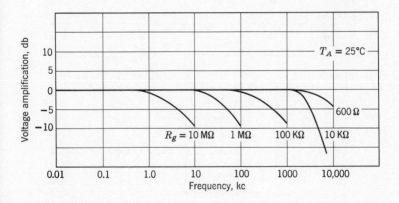

Fig. 21.15 Frequency response of circuit of Fig. 21.14, as function of generator impedance.

40-db High-Input-Impedance Amplifier. Figure 21.16 shows the circuit of an amplifier having an input impedance of 30 megohms,

232 TRANSISTOR APPLICATIONS

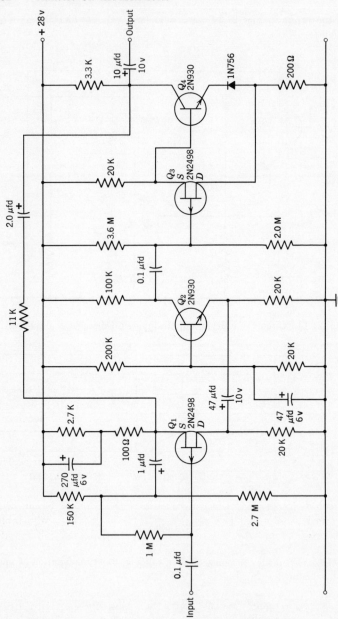

Fig. 21.16 Circuit of 40-db high-input-impedance amplifier.

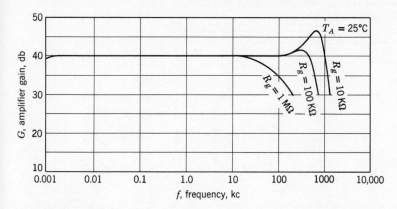

Fig. 21.17 Frequency response of circuit of Fig. 21.16, showing effect of generator impedance.

shunted by 8 pf. Gain is 40 db ± 0.5 db from −55°C to +125°C. The high input impedance is obtained by use of bootstrapping of Q_1. The bias system shown in Fig. 21.5c is employed to permit as much as a 3-to-1 change in $I_{D(\text{on})}$ with no greater than about 12% change in drain current. Wide bandwidth is obtained by operating Q_2 in the grounded-base configuration to reduce the Miller capacitance effect of the FET at high frequencies. By using a FET for Q_3, it is possible to use a large load resistor for Q_2 and thus obtain high voltage amplification.

Figure 21.17 shows the frequency response. Broad-band noise figure varies from 6 db with 10,000-ohm generator impedance to about 1.3 db at 200,000 ohms, and back up to about 3 db at 10 megohms. Average noise figure is about 3 db over a generator impedance range of 50,000 ohms to 5 megohms.

The 60-db Low-Noise Amplifier. Figure 21.18 shows the circuit of a high-gain low-noise amplifier using three cascaded FETs. Adjustable gain is provided up to 60 db ± 0.5 db from −55°C to +125°C. Input impedance is 10 megohms, shunted by 15 pf.

Figure 21.19 shows the frequency response of this amplifier and the effect of generator impedance.

Figure 21.20 shows the equivalent noise input voltage as a function of frequency and Table 21.2 gives the 3-db bandwidth and broad-band noise figure for various values of generator impedance.

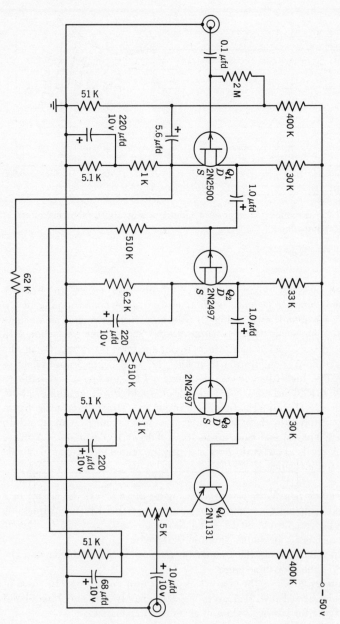

Fig. 21.18 Circuit of 60-db low-noise amplifier.

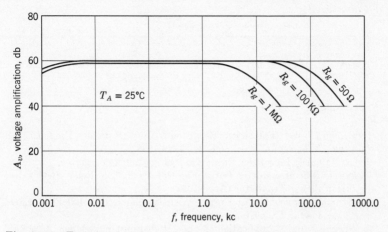

Fig. 21.19 Frequency response of circuit of Fig. 21.18, showing effect of generator impedance.

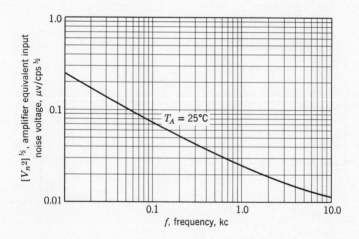

Fig. 21.20 Equivalent input noise voltage versus frequency for amplifier of Fig. 21.18.

Table 21.2

	$R_g = 0$	$R_g = 10$ KΩ	$R_g = 50$ KΩ	$R_g = 100$ KΩ	$R_g = 1$ MΩ
3-db Bandwidth	5 cps–100 kc	5 cps–100 kc	5 cps–60 kc	5 cps–40 kc	5 cps–5 kc
Broad-band noise figure	Equiv. input noise = 3.0 μv	1.5 db	1.2 db	1.0 db	1.0 db

REFERENCES

1. Shockley, W., Transistor Electronics: Imperfections, Unipolar and Analog Transistors, *Proc. IRE*, **40,** 1289 (1952).
2. Crystalonics, Inc., Field Effect Transistor Application Notes.
3. Fairchild Semiconductor Corp., *Fan-Out No. 109*, FSP-400 Field Effect Silicon Planar Transistor (April 1962).
4. Fairchild Semiconductor Corp., Specifications on FSP-400 Field Effect Silicon Planar Transistor.
5. Siliconix, Inc., Application Note, High-Input-Impedance UNIFET Amplifiers (Feb. 1963).
6. Texas Instruments Inc., *Transistor Circuit Design*, McGraw-Hill, 1963, pp. 497–552.

22
Logic Circuits

22.1 Introduction

The subject of logic applications of semiconductor devices is extremely complex, and complete books have been written on this phase of transistor applications alone. It is obviously impossible, therefore, to attempt an all-inclusive treatment within the confines of this chapter. The treatment will be restricted, therefore, to a presentation of the fundamental principles such that the reader may obtain a basic understanding of the subject. For those who wish to explore the subject in depth a number of excellent sources are listed in the references at the end of the chapter.

22.2 The Language of Logic

Logic operations, like most technologies, have fostered a specialized language to indicate the various logic operations, although most of the terms have somewhat similar connotations in everyday usage. In general logic operations consist of combining various inputs to obtain specified outputs, i.e., there is a specific set of rules indicating under which conditions outputs will result. Furthermore, logic operations are binary, i.e., only two conditions are normally encountered, which may have any number of expressions: yes or no, true or false, present or absent, 1 or 0, on or off, etc. It is convenient to use the notations 1 or 0 to indicate the binary states, remembering that these numbers do not necessarily indicate actual voltages but merely their presence or absence.

For example, the actual voltage levels in a logic system may be -2 and $+5$ v, depending on the states of the various switching devices. We can say that -2 v corresponds to the 0 state and $+5$ v to the 1 state, or vice versa, as we wish, as long as we are consistent.

A truth table is a listing of the various conditions pertaining to the logic system. For example, consider a system having three inputs and one output, and having the requirement that the output shall be in the "on" or 1 state when, and only when, two or more of the three inputs are simultaneously "on." This is the familiar two-out-of-three coincidence arrangement, so useful in high-reliability systems. A truth table can be constructed as follows:

	A	B	C	D
(1)	0	0	0	0
(2)	1	0	0	0
(3)	0	1	0	0
(4)	0	0	1	0
(5)	1	1	0	1
(6)	0	1	1	1
(7)	1	0	1	1
(8)	1	1	1	1

The above can be read as: In condition 1, signals are absent on all the inputs, hence there should be no output D. On the next three signal is present on only one input, again not satisfying the requirement, and no output results. In the next three cases two inputs are present and in the last all three are present, under these four conditions output should result, hence $D = 1$. Construction of such a truth table is a very helpful preliminary to any logic design.

Boolean algebra provides a very useful tool for expressing such relationships as the above in mathematical form, and, even more important, for simplifying the resultant equation to its simplest form before proceeding to the step of designing the logic. In this method, a number of familiar algebraic symbols and terms are used in a rather unfamiliar manner. The usual multiplication signs, $\times$, $\cdot$, ( ), etc., now imply the operation "and," while the addition sign $+$ indicates the operation "or." Other symbols used are the overhead bar $^-$ to indicate negation or "not" (implying a specified *absence* rather than *presence* of a signal). Other forms of logic symbols are frequently encountered, but the above symbols will be used herein.

By using this symbology, we can express any relationship, as for example the above two-out-of-three coincidence. Using the conjunction of two or more symbols to denote the operation "and" and the + for "or," the foregoing truth table can be written thus:

$$AB\bar{C} + \bar{A}BC + A\bar{B}C + ABC = 1 \qquad (22.1)$$

This is read as "A and B but *not* C, or *not* A and B and C, or etc." In other words, each of the above three-letter combinations corresponds to one of the four conditions which is supposed to produce an output 1. These are the only combinations which will produce this output, according to our originally stated conditions.

There are a number of rules of Boolean algebra which permit simplification of an expression, such as that in eq. 22.1. These are listed in the following:

$$A + B = B + A \qquad (22.2)$$

$$AB = BA \qquad (22.3)$$

$$(A + B) + C = A + (B + C) \qquad (22.4)$$

$$(AB)C = A(BC) \qquad (22.5)$$

$$AB + AC = A(B + C) \qquad (22.6)$$

$$A + BC = (A + B)(A + C) \qquad (22.7)$$

$$A\bar{A} = 0 \qquad (22.8)$$

$$A + \bar{A} = 1 \qquad (22.9)$$

$$\bar{\bar{A}} = A \qquad (22.10)$$

$$\overline{ABC} = \bar{A} + \bar{B} + \bar{C} \qquad (22.11)$$

$$\overline{A + B + C} = \bar{A}\bar{B}\bar{C} \qquad (22.12)$$

$$AB + \bar{A}B + A\bar{B} + \overline{AB} = 1 \qquad (22.13)$$

$$(A + B)(\bar{A} + B)(A + \bar{B})(\bar{A} + \bar{B}) = 0 \qquad (22.14)$$

$$A\bar{B} + AB = A \qquad (22.15)$$

$$A + AB = A(1 + B) = A \qquad (22.16)$$

$$A + \bar{A}B = (A + \bar{A})(A + B) = A + B \qquad (22.17)$$

Many of these equations seem obvious, e.g., eqs. 22.2 and 22.3, but this simplicity can be deceiving, as shown by eq. 22.7, for example, where the logic equivalence is certainly unlike the usual arithmetic solution.

Let us now illustrate the application of some of these rules to our two-out-of-three problem. We can re-express eq. 22.1 thus:

$$AB(\bar{C} + C) + \bar{A}BC + A\bar{B}C = 1 \qquad (22.18)$$

Since $\bar{C} + C = 1$, this becomes

$$AB + \bar{A}BC + A\bar{B}C = AB + C(\bar{A}B + A\bar{B}) = 1 \qquad (22.19)$$

The quantity inside the parentheses $\bar{A}B + A\bar{B}$ is the "exclusive or," meaning not only that one quantity must be present but also that the other must be absent. In this example, however, since we also have the condition that we can obtain an output with the condition AB, the "exclusive or" can be modified to a simple "or," and the equation finally simplifies to:

$$AB + C(A + B) = 1 \qquad (22.20)$$

This can be read that an output will occur if we have inputs on both A and B, *or* on C and $(A$ *or* $B)$. Figure 22.1 illustrates how this simple relationship could be achieved, using semicircles to indicate the logic circuits, with an ampersand indicating the operation "and" and the inverted caret signifying "or." The results of each logic operation are indicated on the output of each element, and the final combination is obvious.

When any solution to a logic problem has been achieved, a check

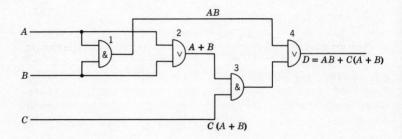

Fig. 22.1 Basic logic scheme for two-out-of-three coincidence.

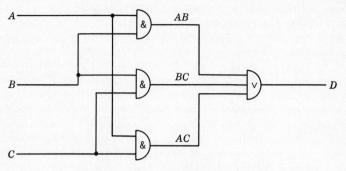

Fig. 22.2 Use of three-input "or" gate.

should be performed by inserting all the original conditions into the solution and making sure that all the specified conditions produce the desired output and no others. For example, in the circuit of Fig. 22.1, if we take condition 2 of our original truth table, i.e., only A is present, we find that block 1 has no output because, being an "and" block, it requires inputs on both terminals. Similar analysis indicates that all other blocks have no outputs. Performing this test on all eight conditions will reveal that this system complies with the original requirement, hence is a solution.

Another possible solution employs multiple gates rather then merely two-input gates. Figure 22.2 illustrates one solution to our example using a three-input "or" gate.

22.3 Logic Elements

There are a number of elementary function blocks in any logic system. These perform the functions described above, namely, "and" and "or," also additional functions, "not," "exclusive or," "nand (not-and)," "nor," and special functions such as delay, memory, etc. Also of considerable utility in logic operations are the flip-flop and set-reset counter. These were described in Chapter 18 and will not be treated here further. The remainder of this chapter will be devoted to a description of the circuits which perform the above logical functions, and to the limitations imposed by such considerations as tolerances, fan-in, and fan-out, etc.

22.4 Types of Logic Circuits

A considerable number of types of logic connections and combinations have been introduced in the past several years. These include direct-coupled transistor logic (DCTL); transistor-resistor logic (TRL); transistor-diode logic (TDL); emitter-coupled logic (ECL or ECTL) and various modifications of these basic forms. Some circuits operate with the transistors going into saturation, in others the saturated state is avoided. Some circuits use "speedup" capacitors to increase switching speed, although in some types this is restricted by such factors as cross talk or noise. Different types have greater or less ability to handle large fan-in or fan-out requirements (fan-in is the number of inputs to a logic block from separate blocks, fan-out is the number of separate outputs).

DCTL circuits are, of course, the simplest fundamentally. However, since they incorporate no coupling resistors or diodes by their very nature, the requirements imposed on the transistors are quite severe. Inequalities of V_{EB}, for example, can cause one transistor to take most of the available base current. Limitations are imposed on the types of transistors which can be used for this type of logic, since the saturation voltage of the driving transistor becomes the input voltage to the driven transistor. DCTL circuits are more feasible in integrated designs because of the greater potential uniformity when all units are made from the same chip.

TRL circuitry is used when speed and maximum fan-in and fan-out are not required, because of its relative simplicity, although, of course, it is not as simple as DCTL. TDL circuits provide for the greatest fan-in and fan-out at the expense of complexity. TDL circuits frequently employ either the "nand" configuration or the "nor." The "nand" circuitry is faster, "nor" circuitry is capable of higher fan-in and fan-out. Generally, if speed is not the major requirement, TDL "nor" circuitry is more economical. ECTL circuits provide much of the simplicity of the DCTL circuits, permit greater fan-in and fan-out, and are ideally suited to integrated circuitry.

Typical examples of the preceding forms of logic will be presented in the following section.

22.5 DCTL Logic

Figure 22.3 illustrates two versions of DCTL [in version (b) the circuit is actually modified DCTL, by virtue of the base resistors added to

LOGIC CIRCUITS 243

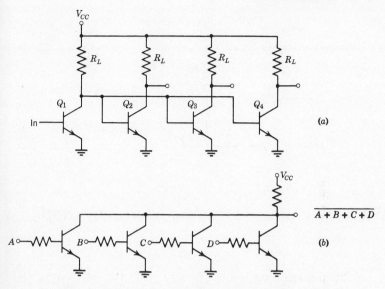

Fig. 22.3 Examples of direct-coupled transistor logic (DCTL).

reduce the current-hogging problem]. In **Fig. 22.3a**, transistor Q_1 is the input transistor and drives the other three by direct connection to their bases. In (b) separate inputs are applied to all four transistors, which have one common output. Version (a) is a single-input non-inverting gate with a fan-out of three. Version (b) is a "nor" gate with fan-in of four. As indicated by the logic equation, all four inputs must be absent for an output to be present.

22.6 TRL Logic

Figure 22.4 shows the basic circuit for the transistor-resistor logic block. There is a fan-in of M and fan-out of N, as shown. A positive input to any of the M inputs will saturate the n-p-n transistor, and the collector will be near ground (binary 0). To produce a binary 1 on the output all inputs must be at 0 level. Thus the equation of this circuit is $\overline{A + B + C}$ and it is a "nor" circuit. The above is based on V_{BB} being a negative voltage to hold the transistor in cutoff until a positive input is applied.

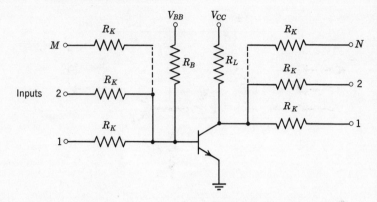

Fig. 22.4 Basic circuit of transistor-resistor logic block (TRL).

Additional transistors can be added to invert the output from TRL gates of the type shown in Fig. 22.4 to produce complementary functions. Thus, if a second *n-p-n* transistor is connected so that the output of the transistor of Fig. 22.4 is connected to its base, the output of the second transistor will be the inverse of the above, or $A + B + C$, and the circuit thus becomes an "and" gate.

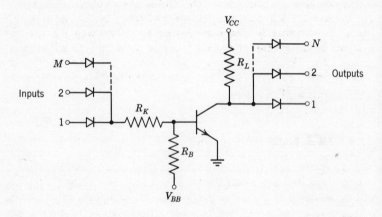

Fig. 22.5 Circuit of transistor-diode (TDL) "nor" gate.

LOGIC CIRCUITS 245

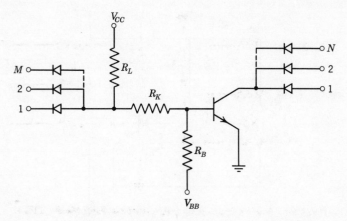

Fig. 22.6 Circuit of transistor-diode (TDL) "nand" gate.

22.7 TDL Logic

Figure 22.5 shows the circuit of a transistor-diode "nor" circuit. As before, a positive input on any of the M inputs will cause the collector voltage to bottom, and the output is therefore a logic 0, unless all inputs are 0, whereupon the output is 1. As before, adding a second transistor to provide inversion will convert this arrangement to an "and" gate.

The circuit of Fig. 22.5 can be converted to a "nand" (*not A and B and C*) by reversing the diodes and moving the load resistor R_L from the collector to the junction of R_K and the input diodes. This is illustrated in Fig. 22.6. The load resistor and the diodes perform the "and" operation on the inputs, and the transistor inverts the signal, resulting in $\overline{ABC}$ for the logic equation.

22.8 ECL Logic

Figure 22.7 shows the circuit of an emitter-coupled "and" gate. V_{BB} can be either a negative bias or a clock signal. In the first case the reference voltage V_{ref} is set midway between V_{BE} and $V_{CE(\text{sat})}$ and is typically about $+0.4$ v. If all inputs to the gate are at $+0.2$ v, the reference transistor Q_R will conduct, causing the output to be near ground or slightly negative. If one of the input transistors has an

246   TRANSISTOR APPLICATIONS

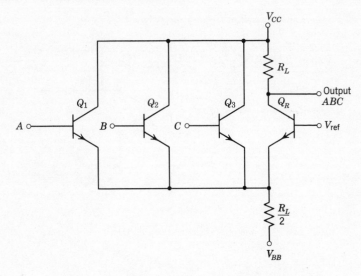

Fig. 22.7   Circuit of emitter-coupled (ECL) "and" gate.

input of $+0.7$ v, that transistor will take most of the current flowing through the emitter resistor and the reference transistor will turn off.

Figure 22.8 illustrates the application of the ECL principle to an "or" gate, and operates according to the same principle as the circuit of Fig. 22.7.

22.9   Tolerance Considerations in Logic Circuit Design

As the fan-in and fan-out numbers go up the available voltage swings become less because of increased loading. Other factors which assume increasing importance are the production variations between such transistor parameters as V_{BE} and $V_{CE(\text{sat})}$ and the variations in supply voltages. Thus, the permissible fan-in and fan-out become a function of these variables.

The relationships between the above variables and the bias and coupling resistors are quite complex. To illustrate this the equations for the TDL "nand" circuit of Fig. 22.6 are given below.

The equation for the "on" condition:

$$R_B = \frac{V_{BB} + V_{BE(\text{on})}}{\dfrac{V_{CC} - R_L M[I_{CBO} + (N-1)I_{DO}] - V_{BE(\text{on})}}{R_L + R_K} - \dfrac{N}{h_{FE}}\left[\dfrac{V_{CC} - (V_{CE(\text{sat})} + V_D)}{R_L} + (M-1)I_{DO} - \dfrac{V_{CE(\text{sat})} + V_D + V_{BE(\text{off})}}{R_K}\right]} \quad (22.21)$$

For the OFF condition:

$$\bar{R}_B = \frac{V_{BB} - V_{BE(\text{off})}}{[(\bar{V}_D + \bar{V}_{CE(\text{sat})} + V_{BE(\text{off})})]/R_K] + I_{CBO}} \quad (22.22)$$

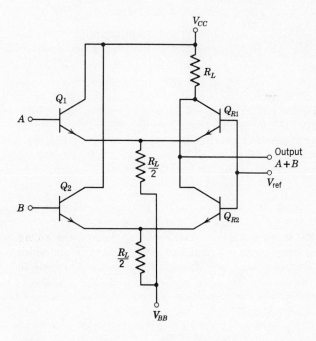

Fig. 22.8 Circuit of emitter-coupled (ECL) "or" gate.

In the above equations an overline indicates a maximum value, an underline indicates a minimum value, magnitudes only are used, M = number of inputs, N = number of outputs, and $V_{BE(\text{off})}$ is assumed to reverse bias the transistor. The above equations give the minimum and maximum values of R_B required to insure turnon and turnoff. If these values are obtained for a range of values of the load resistor R_K and are plotted, the area bounded by the two curves will indicate the zone of permissible performance.

REFERENCES

1. Hunter, L. P. (Ed.), *Handbook of Semiconductor Electronics*, McGraw-Hill, New York, 1962, pp. 15-37–15-83.
2. Hurley, R. B., *Transistor Logic Circuits*, Wiley, New York, 1961.
3. Kvamme, E. F., Microelectronics Using General Electric Emitter-Coupled Logic Operators, *G. E. Application Note 90.80* (Aug. 1962).
4. Shea, R. F. (Ed.), *Transistor Circuit Engineering*, Wiley, New York, 1957, pp. 317–324.
5. General Electric Transistor Manual, Sixth edition, pp. 175–189.
6. Texas Instruments, Inc., *Transistor Logic Design* (Sept. 1962).

23

Integrated Circuits

23.1 Introduction

Integrated circuits, the fabrication of complete circuit assemblies from the basic semiconductor chip, shows strong indications of being the next major advance in the electronics picture. In fact, industry surveys indicate that, by 1968, if not sooner, integrated circuits will equal in dollar volume the entire remaining transistor sales. That this can be so is readily appreciated from the basic advantages of integrated circuits: (1) Since all components are made at the same time, and by essentially identical processes, uniformity within any one assembly should be good; (2) by the same token, since a large number of items are made at once, the total cost should ultimately be far under the cost of assembling the same circuit from discrete components; (3) reliability should be high because of great reduction in the number of handmade interconnections. The growing interest in this subject is, therefore, very understandable.

Integrated circuits are one form of the more inclusive family of microcircuits, which also includes such techniques as miniature pellet assemblies and utilization of thin film techniques. In some cases combinations of these techniques have been desirable, combining, for example, the cost reduction possible with integrated assemblies of active components with the greater uniformity and lower tolerances possible with thin-film passive components. In this chapter, however, attention will be focused on the all-in-one integrated assemblies.

23.2 Types of Integrated Circuits

Initially, integrated assemblies were restricted to combinations of two or more transistors in one package, formed from the same substrate.

As the art advanced, static components, e.g., resistors, were added, and the scope of applications was broadened. The first composite circuits were designed, understandably, for switching operation, i.e., logic, so that uniformity of the components was less stringent than for linear applications. As the ability to maintain close tolerances of components improved, excursions into the linear field of application became more frequent, until currently a number of manufacturers are marketing a wide variety of digital and linear combinations, and the day is rapidly approaching when the designer will have at his disposal sufficient building blocks to permit synthesis of practically any design, with the final product occupying the space originally taken possibly by only one subassembly.

With respect to this ability of the designer to use the full scope of his ingenuity, there is a conflict between the desire to standardize on a reasonable minimum of integrated combinations in order to permit large-quantity production, and thereby reduce cost and improve reliability, and the usual desire on the part of the individual engineer to design something that no one else has thought of. The result is that the available types of integrated designs range from standard blocks to complete custom-built assemblies, with, of course, a corresponding divergence in cost. In between there are a number of compromises, and considerable ingenuity is being exercised to devise means to combine large-quantity production savings with flexibility. Two such compromises of considerable interest are the General Electric Matrix approach and the Texas Instrument Master Slice, which will be described later in this chapter. In the following sections the basic elements of integrated circuit fabrication will be described briefly, and typical examples of the various circuits which have been integrated will be discussed. Also, some of the problems involved in the utilization of this new tool will be touched on, as a guide to the potential user.

23.3 Integrated Circuit Fabrication

Figures 23.1 and 23.2 illustrate the construction of the two basic elements of integrated circuits—the transistor and the resistor. The former can also be used to provide the diode and the capacitor, utilizing the collector depletion layer of a transistor as an effective capacitor. In constructing the above elements, a basic substrate (in this case p-type silicon) is used, upon which all the elements are constructed by the usual transistor fabrication processes, involving epitaxial layer growth, diffusion, passivation and alloying, also etching to clear the various selected

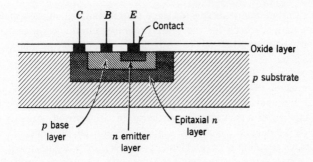

Fig. 23.1 Basic construction of integrated circuit transistor.

areas for successive operations, such as deposition of the ohmic contacts to the various elements. In fabricating a transistor, for example, the n-type epitaxial layer is first produced within the p-type substrate. The silicon dioxide layer which has formed on the surface is etched away selectively and a p-type layer is formed within the n-layer by diffusion. Finally, another n-type layer is formed within the p-type layer, and a final etching exposes the areas to which conducting connections are then made.

In forming the resistor, a similar process is followed, although with fewer steps, and the resistor body is formed by the epitaxial n-type layer within the p-type substrate, with connections being made to the ends of the n-layer. By proper control of the width, length and depth of the n-type strips the desired values of resistance may be obtained.

Since the substrate is the same in both constructions, it becomes

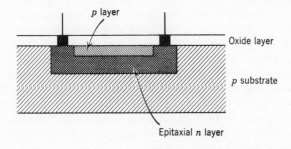

Fig. 23.2 Basic construction of integrated circuit resistor.

obvious that components such as these may be built upon the same chip, thus forming the basis for the total integrated circuit. Use is made of the body of the substrate to provide isolation between the various components.

Inspection of the Figs. 23.1 and 23.2 will also indicate one of the basic limitations of integrated circuits. Since, in effect, the elements are created by constructing transistors within a surrounding substrate, which is in itself a semiconductor, there are also created parasitic semiconductor elements which can shunt the desired elements or cause undesirable coupling between elements unless proper cognizance is taken of this effect in the design. For example, the resistor of Fig. 23.2 is in reality a distributed type of resistor, and there is a distributed p-n-p transistor existing between the p-type isolation layer and the substrate. If the p-n junction of the resistor and the epitaxial layer is forward-biased and the substrate is at a lower voltage, the transistor will be in its active region of operation, with a beta on the order of 0.5–5, hence it can have considerable deleterious effect. To avoid this effect, the substrate should be at such a potential that the internal transistor cannot become active. Similar parasitic elements exist in the form of diodes across or in series with various elements. There are also capacitances between the various layers which can provide shunt paths to affect seriously high-frequency performance. The circuit designer must work closely with the fabricator of the integrated circuits if the ultimate benefit is to be obtained from their use.

23.4 Multiple-Transistor Units

As mentioned in the introduction, the first integrated units were assemblies of two or more transistors, formed from the same chip, and connected in a number of configurations for various specific purposes. Figure 23.3 illustrates one of the simplest of these configurations—the two-transistor Darlington arrangement. It will be recalled that this arrangement has the properties of a supertransistor having a beta derived from the product of the betas of the two component transistors. Such items are manufactured by quite a few manufacturers, e.g., the Westinghouse type WM-1110 functional block, the Fairchild type FSP-4, and the Honeywell MHM1001 and 1101 among others. Some of these units are encased in hermetically sealed flat packages, some in standard TO-5 or TO-18 transistor housings.

Another form of dual-transistor assembly is the differential unit, such as the types 2N2060 and 2N2480, intended for use as the input stage for a d-c differential amplifier. By virtue of the two units being on the same chip, their drifts with temperature will be similar, hence the differential drift will be minimized; similarly, the two units are much more likely to have similar V_{BE} values if mounted on the same chip. Still another common two-transistor assembly is used as the chopper in a carrier-type d-c amplifier, again the common assembly insuring maximum compatibility from the standpoint of cancellation of saturation voltages or leakage currents.

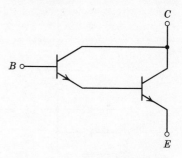

Fig. 23.3 The integrated Darlington unit.

More than two transistors in one package permit considerable versatility on design, particularly in logic circuits. For example, Fairchild has more than 100 transistor-diode multiples, incorporating as many as six transistors or diodes within one TO-5 or TO-18 case, with almost all conceivable combinations of arrangements.

23.5 Integrated Logic Circuits

While multiple assemblies such as described above can be combined with the necessary passive elements to make logic circuits (or, in some cases, e.g., TDL, used completely as logic elements without additional components), usual practice today is to incorporate the passive elements within the same package, in most cases by using the same substrate to form the resistors and diodes as well as the transistors. Typical of this field of application is the RCA DMC 100, the circuit of which is shown in Fig. 23.4. This is a high-speed diode-coupled computer circuit with level shift diodes, designed for use in inverter, positive "nand" or negative "nor" gates, flip-flops, and shift register applications. As shown in Fig. 23.4, the circuit uses diodes as inputs, level-shift diodes as coupling to the transistor base, and a high-speed transistor for switching. It can accommodate as many as 15 outputs by use of external diodes, has a typical rise time of 11 ns, fall time of 10 ns, and stage delay of 7 ns, and can be used to count at rates on the order of 10–20 Mc.

A similar circuit, in dual, is shown in Fig. 23.5, the Westinghouse-type WM-2101 dual-"nand" gate. Each gate has a maximum fan-in of 3

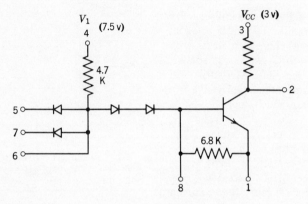

Fig. 23.4 High-speed diode-coupled computer circuit, with level shift diodes.

and fan-out of 5. Typical power dissipation is 10 mw. Time to switch on is 15–16 ns, to switch off 95–65 ns, depending on fan-out, and propagation delay is 55–40 ns.

Figure 23.6 shows the circuit of the General Electric P322 "and" gate, which is typical of the number of emitter-coupled logic circuits (ECTL) possible in this series. As described in Chapter 22, this type of logic block has simplicity comparable to DCTL, yet is less subject to the limitations of that technique with respect to current-hogging or nonuniformity of transistor characteristics.

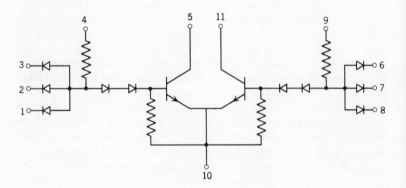

Fig. 23.5 Dual "nand" gate circuit.

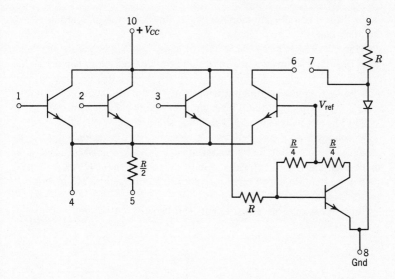

Fig. 23.6 Circuit of emitter-coupled "and" gate.

23.6 Linear Integrated Circuits

Although, as stated above, the initial integrated circuits were designed primarily for digital applications, a growing number of linear applications are yielding to the integrated approach. Two such applications are the differential d-c amplifier and the high-frequency amplifier.

Figure 23.7 shows the circuit of a linear differential amplifier which has been developed in experimental form, with variations, by G.E. and T.I. Basically it consists of a dual differential amplifier Q_1, Q_2 and a constant-current source Q_5. Current from this source can be adjusted to a desired value by inserting an external resistor between terminals 7 and 8, the total emitter resistance in conjunction with the voltage of the breakdown diode D determining the value of the current. Input signals are applied to terminals 10 and 4, and differential outputs are taken from 12 and 2. In a further modification a p-n-p transistor is added to the circuit, with emitter and base connected to terminals 12 and 2, and collector going to the negative terminal 7 through a load resistor. This converts the push-pull output to single-ended, and also permits adjustment so that output is zero for equal inputs to the two input terminals.

By means of external circuit additions, this basic d-c circuit can be made into a bistable, a monostable, a flip-flop, and numerous other modifications.

Figure 23.8 shows the basic circuit used by Westinghouse in their series of high-frequency amplifiers. This circuit is used as shown in the type WM-1106 video amplifier, of which there are four versions, having 3-db down frequencies ranging from 6 Mc to 12 Mc, typically, a minimum power gain of 20 db at 6 Mc, input resistance of 300 ohms, shunted by 230 pf, output resistance 1300 ohms, shunted by 18 pf.

By connecting an external tuning element, such as a lead-zirconate-titanate ceramic filter, between terminals 6 and 1, the circuit becomes an i-f amplifier, with typically a gain of greater than 30 db at a frequency up to 6 Mc, input resistance 350 ohms, shunted by 215 pf, output resistance 200 ohms.

Similarly, a variable-tuned circuit can be substituted for the ceramic

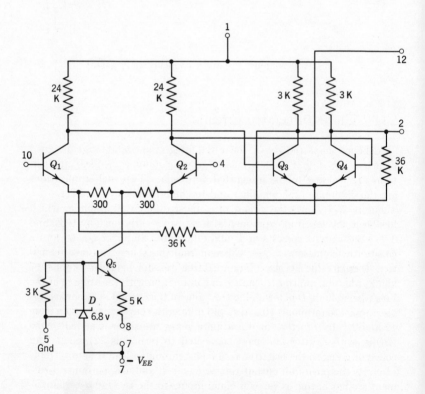

Fig. 23.7 Circuit of linear differential amplifier.

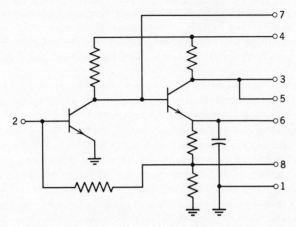

Fig. 23.8 Basic circuit of integrated high-frequency amplifier.

unit and the device becomes a tuned r-f amplifier, also capable of up to 30 db gain at frequencies up to 6 Mc.

By proper combination of these functional blocks, together with a similar oscillator-mixer unit, a complete receiver can be synthesized. There is also a series of audio modules, combining field-effect transistors with conventional transistors, so that output power up to 3 w, single-ended, 6.25 w, push-pull class B, may be obtained.

Texas Instruments has also recently announced a series of linear circuits, incorporating many of the above features, in their 52 series.

23.7 Versatile Custom-Built Assemblies

General Electric and Texas Instruments, among others, have developed basic designs which permit large-quantity production of the basic elements, while at the same time permitting the user to modify available designs with a considerable degree of freedom. This is accomplished by assembling on one unit a large number of transistor elements and resistor elements, which can then be connected together at the customers discretion by applying a connective pattern of his own design. Thus, the basic matrix is standardized and the customer is only required to pay the relatively modest cost of making the pattern for his specific configuration. This technique makes possible very rapid delivery of experimental models, once the pattern is finished.

Figures 23.9 and 23.10 show the basic elements of this matrix approach as employed by G.E.,—the transistor and resistor respectively. Some concept of the size of these elements can be obtained from the fact that there are 1100 transistors and 4000 resistors on a silicon wafer about 1 inch in diameter. (This does not imply, of course, that this number of units would be incorporated into one circuit, but gives an idea of the number of individual complete circuits which can be obtained from one wafer, all essentially alike.)

The versatility of this design can be realized from the following facts: (1) The transistor can be used as an n-p-n unit by using A as the emitter, B as the base, and C as the collector; (2) it will provide a low-leakage high-breakdown diode between B and C; (3) a zener diode (in reverse direction) or reference diode (in forward direction) is obtained between A and B; (4) the collector of a transistor element can be employed as a subterranean transfer point, using the low-resistance connection between any of its C contacts, if conductor crossings are required.

The resistor shown in Fig. 23.10 provides the following resistance values, depending upon connections: 2000 ohms from E to F, 1000 ohms from F to G, 3000 ohms from E to G, and 667 ohms from F to E and G tied together. Thus, it can be seen that this arrangement permits considerable flexibility in circuit arranging.

General Electric has three forms of the basic matrix arrangement, the M1, the original; M2, similar to the M1, but with specific crossovers provided to simplify interconnections; and the M3, which will also incorporate diffused capacitors of about 25 pf.

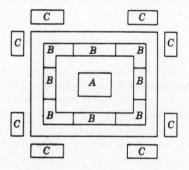

Fig. 23.9 Elements of G.E. integrated transistor. A—emitter contact. B—base contact (any five segments forming a horseshoe shape). C—collector contact.

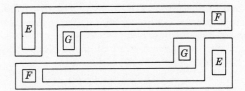

Fig. 23.10 Construction of G.E. integrated resistor.

Texas Instruments also provides a considerable degree of flexibility with their Master Slice concept, which can also be adapted to the customers needs by the deposited interconnections.

23.8 Limitations on Use of Integrated Circuits

The effects of parasitic elements, built-in transistors in parallel with resistor elements, for example, has already been mentioned. Another limitation is the matter of tolerances. Resistors can only be held to $\pm 20\%$, although the ratio between resistors on the same unit can be held to much closer tolerances. It is advantageous, therefore, to design the circuits so that performance is more a matter of relative values or ratios than of absolute values. It is to be anticipated, however, that as the art progresses the ability to hold component tolerances closer than the above will also improve. Further, the yield should materially improve with experience, thus further reducing the cost to little more than that of individual transistors.

REFERENCES

1. Dicken, H., *Parasitic Effects in Integrated Circuits*, Paper *No. FPM 11.1*, presented at the 1963 International Solid-State Circuits Conference, Philadelphia, Pa. (Feb., 1963). Abstract provided in digest of technical papers, pp. 98–99.
2. Dicken, H., *Designing Diffused Integrated Circuit Resistors.*, Motorola Semiconductor Products, Inc., Phoenix, Ariz.
3. Housey, L. J., *Semiconductor Networks for Linear Amplification Using the Switching Mode*, Paper *No. FPM 11.3, ibid.* Abstract on pp. 102–103 of digest.
4. Kvamme, E. F., Microelectronics Using General Electric Emitter-Coupled Logic Operators, *G. E. Application Note 90.80* (Aug. 1962).

5. Lin, H. C., M. J. Geisler, and K. K. Yu, *A Unipolar Transistor Configuration for Integrated Audio Amplifiers*, Paper No. FPM 11.2, ibid. Abstract on pp. 100–101 of digest.
6. Luecke, G., *A Semiconductor Network Multiplex Switch*, Paper No. FPM 11.5, ibid. Abstract on pp. 106–107 of digest.
7. Narud, J. A., W. C. Seelbach and N. Miller, *Relative Merits of Current Mode Logic Microminiaturization*, Paper No. FPM 11.4, ibid. Abstract on pp. 104–105 of digest.
8. Fairchild Semiconductor Corp., *The Inside Story on Fairchild Micrologic*.
9. Texas Instruments, Inc., *Master Slice, Solid Circuit Semiconductor Networks*.

A
Symbols

A	Amplification, amplifier
A_i	Current amplification
A_N	Voltage ratio of oscillator network
A_v	Voltage amplification
A_{vc}	Closed-loop voltage amplification
A_{vo}	Open-loop voltage amplification
a_{ij}	Term in a matrix
B	Base
B_n	Bandwidth over a specified level
B_n'	Normalized bandwidth
BV	Breakdown voltage
BV_{DS}	Breakdown voltage, drain-to-source, field-effect transistor (FET)
b_{fs}	Small-signal common-source forward transfer susceptance (FET)
b_{ij}	Term in b matrix
b_{is}	Small-signal common-source input susceptance (FET)
C	Capacitance
C_{BE}	Base-emitter depletion layer capacitance
C_{be}'	Internal base-emitter capacitance, high-frequency equivalent circuit
C_c	Collector capacitance
C_{cb}'	Collector-internal base capacitance, high-frequency equivalent circuit
C_{ec}	Emitter-collector capacitance, high-frequency equivalent circuit

C_E	Emitter by-pass capacitance
C_i	Input capacitance
C_o	Output capacitance
C_R	Reset-pulse capacitor
C_S	Set-pulse capacitor
C_s	Stray capacitance
C_T	Trigger-pulse capacitor
D	Diode
D_S	Steering diode
D_T	Discharge diode
d	The differential operator
E	Emitter
E	Voltage
e_g	Generator voltage
FET	Field-effect transistor
f	Frequency
f_0	Center frequency of pass band
f_{osc}	Frequency of oscillation
f_{ro}	Tunnel diode resonant cutoff frequency
f_{xo}	Tunnel diode self-resonant frequency
G	Power gain
G_m	Maximum value of power gain
G_t	Transducer gain
g	Conductance
g_{be}'	Internal base-emitter conductance, high-frequency equivalent circuit
g_c	Collector conductance
g_{cb}'	Collector to internal base conductance, high-frequency equivalent circuit
g_{cd}	Collector diffusion conductance
g_d	Tunnel diode negative conductance
g_{ec}	Emitter-collector conductance, high-frequency equivalent circuit
g_{fs}	Small-signal common-source forward transfer conductance (FET)
g_i	Input conductance
g_{ij}	Term in g matrix
g_{is}	Small-signal common-source input conductance (FET)
g_m	Mutual conductance
g_o	Output conductance
h_{11}	Network input impedance, output short-circuited
h_{12}	Network reverse voltage transfer ratio, input open

h_{21}	Network forward current transfer ratio, output short-circuited
h_{22}	Network output admittance, input open
h_{FB}	Ratio of d-c collector current to emitter current
h_{fb}	Forward current transfer ratio, output short-circuited, common-base configuration
$(h_{fb})_0$	Low-frequency value of h_{fb}
h_{fc}	Forward current transfer ratio, output short-circuited, common-collector configuration
h_{FE}	Ratio of d-c collector current to base current
h_{fe}	Forward current transfer ratio, output short-circuited, common-emitter configuration
$(h_{fe})_0$	Low-frequency value of h_{fe}
h_{feI}	Imaginary part of h_{fe}
h_{feR}	Real part of h_{fe}
h_{ib}	Input impedance, output short-circuited, common-base configuration
$(h_{ib})_0$	Low-frequency value of h_{ib}
h_{ic}	Input impedance, output short-circuited, common-collector configuration
h_{ie}	Input impedance, output short-circuited, common-emitter configuration
h_{ieI}	Imaginary part of h_{ie}
h_{ieR}	Real part of h_{ie}
h_{ob}	Output admittance, input open, common-base configuration
$(h_{ob})_0$	Low-frequency value of h_{ob}
h_{oc}	Output admittance, input open, common-collector configuration
h_{oe}	Output admittance, input open, common-emitter configuration
h_{rb}	Reverse voltage transfer ratio, input open, common-base configuration
$(h_{rb})_0$	Low-frequency value of h_{rb}
h_{rc}	Reverse voltage transfer ratio, input open, common-collector configuration
h_{re}	Reverse voltage transfer ratio, input open, common-emitter configuration
I	Current
I_B	D-c base current
I_b	A-c base current
I_{BS}	Base current required to saturate collector current
I_C	D-c collector current

I_c	A-c collector current
I_{CBO}	Saturation current, collector-base diode
I_{CS}	Collector saturation current
I_D	Drain current
$I_{D(on)}$	Drain current at zero bias
I_E	D-c emitter current
I_e	A-c emitter current
I_{EO}	Emitter reverse current, unijunction transistor
I_g	Current generator
I_h	Holding current
I_i	Input current
I_o	Output current
I_S	Switching current
i_B	Instantaneous value of base current
i_C	Instantaneous value of collector current
i_E	Instantaneous value of emitter current
J	Current
j	$\sqrt{-1}$
k	Boltzmanns constant
k	Coefficient of coupling
L	Inductance
L_S	Tunnel diode series inductance
l	Minority carrier diffusion length
M	Fan-in number
M	Mutual inductance
M_N	Impedance ratio of oscillator network
m	Transformer turns ratio
N	Fan-out number
N	Network
NF	Noise figure
$NF_{(opt)}$	Optimum noise figure
n	Bandwidth number
n	Transformer turns ratio
P	Power
P_i	Input power
P_o	Output power
p	Ratio, base width to minority carrier diffusion length
p'	$\partial p / \partial V_{CB'}$
Q	Quality factor of coil
Q	Transistor
Q_0	Unloaded Q of coil
q	Electron charge

SYMBOLS

R	Resistance
R_B	Effective base bias resistance
R_{B1}, R_{B2}	Components of unijunction transistor interbase resistance
R_b	A-c value of base bias resistance network
R_{BB}	Interbase resistance, unijunction transistor
R_C	D-c collector load resistance
R_c	A-c collector load resistance
R_D	External drain resistance
R_E	External d-c emitter resistance
R_e	External a-c emitter resistance
R_f	Feedback resistance
R_G	Effective gate bias resistance
$R_{G(opt)}$	Optimum generator resistance (FET)
R_g	Generator resistance
R_i	Input resistance
R_i	Real part of input impedance
R_{im}	Input impedance for matched condition
R_k	Coupling resistance
R_L	Load resistance
R_l	Load resistance
R_{lm}	Load resistance for matched condition
R_l	Real part of load impedance
R_S	External source resistance
r_b'	Base spreading resistance
r_ϵ	Emitter diffusion resistance
S	Stability factor
S_{IE}	Stability factor, dI_E/dI_{CBO}
S_{IC}	Stability factor, dI_C/dI_{CBO}
S_V	Stability factor, dV_{CB}/dI_{CBO}
S_{V1}	Stability factor, dV_{CB}/dV_{CC}
S_{V2}	Stability factor, dV_{CB}/dV_{EE}
s	Complex frequency
T_A	Ambient temperature
T_J	Junction temperature
t	Time
t_d	Delay time
t_f	Fall time
t_r	Rise time
t_s	Storage time
V	Voltage
V_{BB}	D-c base supply voltage; d-c interbase voltage
V_{BE}	D-c base-emitter voltage

V_{CB}	D-c collector-base voltage
V_{CC}	D-c collector supply voltage
V_{CE}	D-c collector-emitter voltage
$V_{CE(\text{sat})}$	Saturated value of V_{CE}
V_D	Diode voltage
V_{DS}	D-c drain-source voltage
V_{EB}	D-c emitter-base voltage
$V_{E(\text{sat})}$	Emitter saturation voltage, unijunction transistor
V_{EE}	D-c emitter supply voltage
V_{FP}	Tunnel diode forward voltage at current I_P
V_G	D-c gate supply voltage, FET
V_g	Generator voltage
V_{GS}	D-c gate-source voltage
V_h	Holding voltage
V_i	Input voltage
V_n	Input voltage to oscillator network
V_n	Amplifier equivalent noise voltage
V_o	Output voltage
V_P	Peak voltage
V_{ref}	Reference voltage
V_S	Switching voltage
V_{sat}	Saturation voltage
V_V	Valley voltage
v_{EB}	Instantaneous value of emitter-base voltage
W	Base width
x	Loading factor
Y	Admittance
Y_f	Feedback admittance
Y_g	Source admittance
Y_i	Input admittance
Y_l	Load admittance
Y_o	Output admittance
Y_{oe}	Output admittance, common-emitter configuration
y	Admittance
y	Loading factor
y_{11}	Input admittance, output short-circuited
y_{12}	Reverse transfer admittance, input short-circuited
y_{21}	Forward transfer admittance, output short-circuited
y_{22}	Output admittance, input short-circuited
y_{fb}	Forward transfer admittance, output short-circuited, common-base configuration

y_{ib}	Input admittance, output short-circuited, common-base configuration
y_{rb}	Reverse transfer admittance, input short-circuited, common-base configuration
y_{ob}	Output admittance, input short-circuited, common-base configuration
Z	Impedance
Z_E	External emitter impedance
Z_e	External emitter impedance
Z_f	Feedback impedance
Z_g	Generator impedance
Z_i	Input impedance
Z_{ie}	Input impedance, common-emitter configuration
Z_l	Load impedance
Z_o	Output impedance
Z_{oe}	Output impedance, common-emitter configuration
z_{11}	Input impedance, output open
z_{12}	Reverse transfer impedance, input open
z_{21}	Forward transfer impedance, output open
z_{22}	Output impedance, input open
z_{fb}	Forward transfer impedance, output open, common-base configuration
z_{ib}	Input impedance, output open, common-base configuration
z_{rb}	Reverse transfer impedance, input open, common-base configuration
z_{ob}	Output impedance, input open, common-base configuration

GREEK SYMBOLS

α_{b0}	Common-base low-frequency short-circuit transfer ratio
β	Common-emitter low-frequency short-circuit transfer ratio
Δ	Determinant
Δ^a	Determinant of a matrix
Δ^b	Determinant of b matrix
Δ^g	Determinant of g matrix
Δ^h	Determinant of h matrix
Δ^y	Determinant of y matrix
Δ^z	Determinant of z matrix
∂	Partial differential operator
η	Efficiency
η	Intrinsic stand-off ratio, unijunction transistor
η	Power transfer efficiency

η_{max}	Maximum power transfer efficiency
ρ	Selectivity factor
θ	Shape factor of coupled tuned circuit
μ_0	Low-frequency reverse voltage transfer ratio
τ_a	Life time in active region
τ_b	Life time in saturated region
ω	Angular frequency
ω_{ab}	Angular cutoff frequency, common-base configuration.

Index

A matrix, 4
a parameters, 4
Amplification, current, 68, 69, 70
    of a cascaded pair, 78
    voltage, 68, 69, 70
    of a cascaded pair, 78
Amplifier, carrier-modulation, 106
    chopper, 104, 253
    class-A power, 117
    class-B power, 121
    complementary class B, 127
    d-c, 97
    differential, 101, 252, 255
    direct-coupled, 97
    distributed, 166
    field-effect transistor, 108, 229
    high-frequency, 161, 256
    high-input-impedance, 231
    logarithmic, 107
    low-noise, 233
    low-output-impedance, 128
    RC-coupled, 81, 114
    transformer-coupled, 112
    tunnel diode, 211
    unity gain, 229
    video, 155
    wide-band, 155
    zero-stabilized, 107
and gate, 244, 245, 254
Astable multivibrator, 178, 214

B matrix, 4
b parameters, 4

Backward diode, 218
Bandwidth, of a tuned amplifier, 141
    normalized, 142
Base spreading resistance, 44
Bias, equations for, 56, 57, 58
    single-battery, 55
    two-battery, 55
Bias stability factors, 57, 58
Bistable multivibrator, 183
Boolean algebra, 238
Bridge, varactor, 106

Capacitance, collector, 44
    emitter, 196
Capacitor, coupling, effect of, 85
    emitter by-pass, effect of, 86, 87
Carrier-modulation amplifier, 106
Characteristics, field-effect transistor, 40
    input, 27
    output, 28
Charge, base, 195
    collector, 196
    emitter, 194
Charge control, 192
Chopper-type amplifier, 104, 253
Clapp oscillator, 177
Class A, distortion, 115
    operation, 117
Class B, bias, 127
    distortion, 124, 125, 126
    operation, 121

Collector diffusion conductance, 42
Colpitts oscillator, 171
Common-mode rejection, 103
Compensation, high-frequency, 155
  low-frequency, 161
Complementary class-B amplifier, 127
Complex frequency, 188
Conductance, collector, 44
  collector diffusion, 42
  mutual, 223
Configuration, common-base, 26
  common-collector, 26
  common-emitter, 26
Counter, ring, 207
Coupling, coefficient of, 146
  transitional, 147, 148
Crossover distortion, 125, 126
Current, holding, 204
  peak, 202, 207
  switching, 204
  valley, 202, 207
Current amplification, 68, 69, 70, 78
Cutoff frequency, 47, 208

Darlington pair, 93, 252
  parameters of, 96
D-c amplifiers, 97
Degenerated common-emitter stage, parameters of, 89
Delay time, 194
Demodulator, diode, 105
Determinant interrelations, 8
Determinants of matrices, 6
Differential amplifiers, 101, 252, 255
  with feedback, 104
Diode, backward, 218
  four-layer, 204
  logarithmic, 107
  steering, 184

Diode, tunnel, 207
Diode bridge modulator, 106
Diode-transistor logic, 242, 245, 253
Direct-coupled amplifier, 97
Direct-coupled logic, 242
Direct-drive power amplifier, 128
Dissipation, in class-B amplifiers, 123
Distortion, crossover, 125, 126
  in class-A amplifiers, 115
  in class-B amplifiers, 124, 125, 126
Distributed amplifier, 166
Drain, 222
Double-tuned circuits, shape factor, 146
  transitional coupling, 147, 148

Efficiency, maximum transfer, 144
  of a class-B amplifier, 123
  power transfer, 143
Emitter by-pass capacitor, effect of, 86, 87
Emitter diffusion resistance, 42
Emitter follower, 65
Emitter impedance, effects of, 86, 87
Equivalent circuit, common-base, 66
  common-collector, 67
  common-emitter, 46, 66
  high-frequency, 134

Fall time, 196
Fan-in, 242, 246
Fan-out, 242, 246
Feedback, collector-base, 89
  emitter, 85, 87
  in broadband amplifiers, 163
Field-effect transistor, 221
  characteristics of, 40

Field-effect transistor amplifier, 108
Flip-flop, 183, 206, 216
Four-layer diode, 204
Frequency, cutoff, 47, 208
   self-resonant, 208
   transition, 135
Frequency compensation, high, 155
   low, 161
Frequency divider, 215

G matrix, 4
g parameters, 4
Gain, 68, 69, 70
   maximum, 68, 69, 70
   transducer, 17, 68, 69, 70
   of cascaded stages, 80
Gate, 222

H matrix, 3
h parameters, definition of, 3, 33
   functions of operating point, 36
   functions of temperature, 36
   high-frequency, 47, 132
   ideal transistor, 42
h_{fb}, definition of, 34
h_{fc}, definition of, 34
h_{fe}, definition of, 34
h_{FE}, definition of, 34
h_{ib}, definition of, 33
h_{ic}, definition of, 33
h_{ie}, definition of, 33
h_{ob}, definition of, 34
h_{oc}, definition of, 34
h_{oe}, definition of, 34
h_{rb}, definition of, 34
h_{rc}, definition of, 34
h_{re}, definition of, 34
Hartley oscillator, 174
High-frequency compensation, 155

High-frequency integrated amplifier, 256
Hybrid combination, 215

I_{CBO}, definition of, 27
   variation with temperature, 53
Ideal transistor, 42
I-f amplifier, integrated, 256
Impedance, input, 68, 69, 70
   of cascaded network, 78
   of degenerated CE stage, 70
   output, 68, 69, 70
Input characteristics, 27
Integrated assemblies, 249
Integrated circuit elements, 250
   matrix, 258
   resistor, 251, 259
   transistor, 251, 258
Integrated linear circuits, 255
Integrated logic circuits, 253
Interbase resistance, 201
Interbase voltage, 200
Intrinsic standoff ratio, 201

Laplace transforms, 189
Lifetime, 196
Linearity of class-A amplifiers, 115
   of class-B amplifiers, 124
Logarithmic d-c amplifiers, 107
Logic, direct-coupled transistor, 242
   elements, 241
   emitter-coupled, 242, 245, 254
   integrated circuits, 253
   terms, 237
   transistor-diode, 242, 245, 253
   transistor-resistor, 242, 243
   truth table, 238
Low-frequency compensation, 161

Matrix interrelations, 7
Matrix, transfer, 92

Matrix, transposition, 92
Maximum power gain, 68, 69, 70
Measurement of parameters, 34
Microcircuits, 249
Millman theorem, 20
Modulator, chopper-type, 104
  varactor bridge, 106
Monostable multivibrator, 182, 214
Multiplication of matrices, 12
Multivibrator, astable, 178
  bistable, 183
  monostable, 182
  tunnel diode, 214
Mutual conductance, 223

nand circuit, 242, 245, 253
Negative conductance, 207
Negative resistance, 178, 199
Noise figure, of field-effect transistor, 226
nor circuit, 242, 243, 245, 253

One-shot multivibrator, 182
Operating point, determination of, 58
  variation with temperature, 62
or gate, 246
Oscillator, Clapp, 177
  Colpitts, 171
  Hartley, 174
  phase-shift, 175
  relaxation, 202
  tunnel diode, 211
Output characteristics, 28
Output impedance, 68, 69, 70

Parameters, common-base, 48
  common-collector, 48
  common-emitter, 48
  high-frequency, 47, 132
  measurement of, 34

Parameters, of field-effect transistor, 225
  of ideal transistor, 42, 131
  of real transistor, 44
Parasitic elements in integrated circuits, 252
Peaking, high-frequency, 155
Pellet assemblies, 249
Phase-shift oscillator, 175
Pinch-off, 220
Power gain, 17
Power output, rms, 118
Power transfer efficiency, 143
Pulse generator, 206

Q factor, 140
Quiescent operating point, 58

RC-coupled amplifier, 81, 114
Relaxation oscillator, 202
Resistance, base spreading, 44
  emitter diffusion, 42
  input, 68
  negative, 178, 199
  output, 68
Resistive cutoff frequency, 208
R-f amplifiers, 161, 256
Ring counter, 207
Rise time, 194

Saturation resistance, 28
Saturation voltage, 29
Sawtooth generator, 203, 206
Schmitt trigger, 185
Self-resonant frequency, 208
Shape factor, 146
Shunt feedback, 89
Single-tuned circuits, 142
Source, 222
Source follower, 229
Stability factors, 57, 58
Stabilization, bias, 56

INDEX

Standoff ratio, intrinsic, 201
Steering diode, 184
Storage time, 194

Temperature, effects of, 62
   variation of I_{CBO}, 53
Thin film circuits, 249
Transducer gain, 17, 68
   of cascaded stages, 80
Transfer impedance, 146
Transfer matrix, 92
Transformer coupling, 112
Transient response, common-base configuration, 190
   common-collector configuration, 191
   common-emitter configuration, 191
   large-signal, 192
   small-signal, 187
Transient waveforms, 193
Transistor, unijunction, 199
   unipolar, 221
Transistor chopper, 104, 253
Transition frequency, 135
Transitional coupling, 147, 148
Transposition matrix, 92
Tuned amplifiers, 140
Tunnel diode, 207

Tunnel diode, amplifier, 211
   flip-flop, 216
   frequency divider, 215
   multivibrator, 214
   oscillator, 213
Two-port networks, 1

Unijunction transistor, 199
Unipolar transistor, *see* Field-effect transistor

Varactor bridge modulator, 106
V_{BE}, variation with temperature, 30, 53
Video amplifiers, 155, 255
Voltage, avalanche, 204
   holding, 204
   peak, 202, 207
   switching, 204
   valley, 202, 207

Wide-band amplifiers, 155

Y matrix, 3
y parameters, 3

Z matrix, 2
z parameters, 2
Zero-stabilized amplifier, 107